Seeding Solutions

Volume 1. Policy options for genetic resources:
People, Plants, and Patents revisited

The Crucible II Group

Copublished by
the International Development Research Centre,
the International Plant Genetic Resources Institute
and the Dag Hammarskjöld Foundation

INTERNATIONAL PLANT GENETIC RESOURCES INSTITUTE
Via delle Sette Chiese, 142, 00145 Rome, Italy (http://cgiar.org/ipgri)
ISBN 92-9043-443-0

DAG HAMMARSKJÖLD FOUNDATION
Övre Slottsgatan 2, 753 10 Uppsala, Sweden (http://www.dhf.uu.se)
ISBN 91-85214-27-2

INTERNATIONAL DEVELOPMENT RESEARCH CENTRE
PO Box 8500, Ottawa, ON, Canada K1G 3H9 (http://www.idrc.ca)

Canadian Cataloguing in Publication Data

Crucible II Group

Seeding solutions. Volume 1. Policy for genetic resources (*People, plants, and patents* revisited)

Copublished by the International Plant Genetic Resources Institute (IPGRI), and the Dag Hammarskjöld Foundation (DHF).
Includes bibliographical references.
ISBN 0-88936-926-7

1. Germplasm resources, Plant.
2. Plant varieties — Protection.
3. Plants, Cultivated — Patents.
4. Biological diversity conservation.
5. Patents.
I. International Plant Genetic Resources Institute.
II. Dag Hammarskjöld Foundation.
III. International Development Research Centre (Canada)
IV. Title.
V. Title: Policy options for genetic resources (*People, plants, and patents* revisited)

K1401.C78 2000   333.95'34   C00-980188-X

Printed by Litopixel, Rome, Italy

Seeding Solutions

Volume 1

Funding Organizations for the Crucible II Project
German Federal Ministry for Economic Cooperation and Development/ German Technical Cooperation (BMZ/GTZ, Germany)
Canadian International Development Agency (CIDA, Canada)
Dag Hammarskjöld Foundation (DHF, Sweden)
International Development Research Centre (IDRC, Canada)
Swiss Agency for Development and Cooperation (SDC, Switzerland)
Swedish International Development Cooperation Agency (Sida-SAREC, Sweden)

Partner Organizations
International Plant Genetic Resources Institute (IPGRI, Italy)
Rural Advancement Foundation International (RAFI, Canada)

Contents

Preface
The Crucible Groups

In 1993, in the aftermath of the UN Conference on Environment and Development (UNCED) and in anticipation of the finalization of the General Agreement on Tariffs and Trade (GATT) Uruguay Round of Multilateral Trade Negotiations, a group of 28 individuals from 19 countries met first in Rome and then in Uppsala and Bern. Members of the group came from South and North; from the private and public sectors and from civil society organizations. Some were scientists while others were policy- and opinion-makers or business executives. Although the group, which began referring to themselves as the Crucible Group, held vastly differing views on many controversial issues, they shared a concern for the conservation and enhancement of plant genetic resources and an alarm that decisions were being taken or policies adopted that could imperil the availability of plant genetic resources for world food security and for agricultural development.

In an effort to clarify issues and choices for decision-makers, the Crucible Group agreed to debate the most contentious points among themselves and to prepare a non-consensus report that would simply lay out the best arguments of every side. Following many months of e-mail and face-to-face debate, the Group identified ten areas where no agreement was possible but where they could offer distinct 'viewpoints' that might prove helpful to others. Most of these issues involved intellectual property related to living organisms, the role of the CGIAR (Consultative Group on International Agricultural Research) and the future structure of an international genetic resources conservation and exchange system.

To its own surprise, however, the Group was able to identify 28 recommendations they felt able to offer collectively to policy- and opinion-makers. In June 1994, *People, Plants, and Patents* was released at a seminar hosted by the International Development Research Centre (IDRC) in Ottawa. Following the release of the book, diverse members of the Group followed up their report with seminars and workshops at the Biodiversity Convention meetings in Nairobi, Nassau, Djakarta and Montreal and at United Nations Food and Agriculture Organization (FAO) meetings in Rome and Leipzig. The book was translated into French and Spanish and was widely distributed.

Although the Crucible Group had made no plans to continue, five years after their first full session at the Dag Hammarskjöld Foundation in Uppsala (DHF), many of the same people found themselves together again in the same place, revisiting the same unresolved issues. Although the group that met in 1998 had not entirely planned this, they quickly agreed that there was a need to convene 'Crucible II' and to try to move the international agenda for genetic resources further. Efforts were made to broaden the Group's membership and the style of dialogue was updated to take better account of internet and e-mail text negotiations. By the end of this second Crucible round, in addition to the first Uppsala meeting hosted by DHF, Crucible participants have gathered at large-scale working meetings held in Ottawa, hosted by IDRC; Nairobi, hosted by the African Centre for Technology Studies (ACTS); and Rome, hosted by the International Plant Genetic Resources Institute (IPGRI). Five years after the publication of *People, Plants, and Patents*, this new volume is released; it will be followed by a second volume. Members of the Group are committed to presenting and discussing their ideas and conclusions, to the extent others find this useful, in the months ahead.

The Crucible Group, now even more than in its first round, continues to be a highly diverse gathering of individuals who passionately and respectfully disagree on intellectual property, the rights of farmers, the mechanisms for benefit-sharing, and the appropriate structures for conservation. More than 45 individuals from 25 countries took part in one or more of the face-to-face discussions and exchanged opinions and data electronically. As with the original Crucible members, Crucible II also shares a passion for plant genetic resources and an ever-growing alarm that one of humanity's most vital resources is being threatened or squandered. Beyond this, the members have 'agreed to disagree' and have worked hard and (strange though it may seem) cooperatively to describe their differences without compromise. Will there be a Crucible III? That will depend entirely on how these urgent issues play out in the years ahead.

The Crucible II reports: *Seeding Solutions*

Those familiar with *People, Plants, and Patents** will recall that the book offered a summary of the major issues related to the ownership, conservation and exchange of plant germplasm. By and large, Volume 1 of the Crucible II documents serves a similar purpose. The volume, as succinctly as possible, brings readers up to date on what has changed, cientifically, politically, environmentally, since the first report five years ago. Readers will find — and hopefully benefit from and even enjoy — new 'viewpoint boxes' that summarize the state of the debate as it stood in late 1999. Readers will also be pleased to find surprising areas of agreement in the form of recommendations. From beginning to end, Volume 1 offers policy-makers a clear description of the facts, the fights and the fora relevant to genetic resources. Those new to these issues will also be offered a clear picture of

---

* The full text of the first Crucible book *People, Plants, and Patents* is available free at http://www.idrc.ca

why germplasm is important and how it relates to trade negotiations, intellectual property disputes, and national and international food and health security.

Volume 2 of *Seeding Solutions* does *not* provide the answers to the issues posed in Volume 1. Indeed, some Crucible members might argue that Volume 2 demonstrates the absurdity of trying to solve sociopolitical issues identified in Volume 1 through the application of purely legal mechanisms. Nevertheless, Crucible II's second volume does extract three broad themes arising from the first volume and Crucible members offer a discussion and a description of — but not a prescription for — some of the legal options available to national governments. Broadly speaking, Volume 2 includes legal mechanisms to address three main points: (1) the need to conserve and exchange germplasm for the benefit of present and future generations; (2) the need to encourage innovation in the conservation and enhancement of germplasm; and (3) new options for securing and strengthening the rights and interests of indigenous and rural peoples in their role as creators and conservers of biological diversity. To the extent possible and in the space manageable, Crucible members have provided their 'best effort' in discussing the range of legal mechanisms, both traditional and *sui generis*, available to policy-makers to resolve these needs.

Readers are urged to study both volumes and to examine Volume 1 before launching into the more legalistic debates in Volume 2. Together, the Crucible II Group hopes that you will find these reports helpful to your own understanding of the issues and in your own policy- and opinion-making activities.

The contents of this volume

Part One: Context of this volume has two sections. The first, 'A Wider Lens for Considering Biodiversity and Intellectual Property Issues', provides a wider context for understanding the intellectual property (IP) and biodiversity debate. What are the social, economic and environmental factors that influence intellectual property and biological diversity in the dawn of the new millennium? What are the recent changes in social attitudes and social awareness that affect the biodiversity and IP debate? This chapter outlines the issues of the loss of biological and cultural diversity, global climate change and increased recognition for the role of farmers and indigenous peoples in conserving, developing and using biological diversity. It also examines the changing roles of public and private sector agricultural research.

The second section of 'the context' deals with 'Changes in molecular bioscience'. What has changed in molecular biosciences and how does that influence the way society thinks about, uses and values biodiversity? Biological knowledge is expanding rapidly. By one estimate, the ability to identify and use genetic information is doubling every 12 to 24 months.[*] Scientific and technical breakthroughs at the molecular level are not only changing the practice or interpretation of science, they often have profound implications for society. Among the examples highlighted in this section are mammalian cloning, advances in genomics, the engineering of plants that render second generation seeds sterile, advances in drug research and discovery, and artificial human chromosomes.

Part Two: Outstanding issues scrutinizes three major areas of discussion: 'Access and Exchange, Knowledge, and Innovation'. Who are the players and what are the fora where biological diversity and IP are being discussed? Over the past five years the international community has seen major conventions and legal agreements enter into force that relate to conservation and use of biological diversity and/or the control, ownership of and access to biological materials. Negotiations on biological diversity and intellectual property are taking place in multiple fora with overlapping and sometimes contradictory objectives. However, there is a real danger of losing track of the overarching themes and trends amidst the minutiae of international conventions. This section examines the central policy issues in three wide areas: policies related to germplasm access and exchange; policies linked to knowledge conservation and formation; and policies involving innovation management.

Crucible Group viewpoints and recommendations

The Crucible Group operates on the basis of good faith — producing best effort non-consensus texts. Members of the Group are individuals attending solely in their personal capacities. They have agreed to associate their names with this volume in the belief that the texts represent a helpful contribution to the global discourse on these issues. Members believe that the texts, in sum, offer an accurate representation of the range of opinions in play at this time and that these divergent viewpoints should be addressed. Probably, every member of the Group is in strong disagreement with some general statements and many specific views provided in both volumes.

In the process of debating issues, the Group has often concluded that readers would best be helped by a 'viewpoint box' that would lay out two or three different perspectives on an important issue. The criteria for accepting each viewpoint perspective is that it fairly represents a position held by one or more stakeholders. It is, however, not unusual that a box contains perspectives that are, in fact, not held by any member of the Group. The style of the boxes is intended to represent the stated informal language of those holding the viewpoint and all members of the Group have worked together

[*] Monsanto, Annual Report, 1997. On the internet: http://www.monsanto.com

to ensure that the best arguments are made on behalf of this perspective.

Sometimes the Group has opted to provide both a viewpoint box and a recommendation. This normally occurs when the Group shares a common view which it recognizes is significantly different from the general debate. All the 15 recommendations cited are those of the whole Group although some individuals undoubtedly associate themselves more with some recommendations than others and, in many cases, individuals believe that the recommendations represent the 'lowest common denominator' within the membership. Despite this, the Group has tended to avoid platitudes and to strive for the highest-achievable common denominator.

Acknowledgements
Members and management of the Crucible II Group

Crucible Group members participate in their personal capacity, but they all have backgrounds in sectors or activities that inform their views on many of the issues that are taken up in this publication. Consequently, we have sub-divided the participants/signatories to this text into four categories: each of which reflects the sector in which the participant has spent the majority of his or her time in recent years. We do not, however, provide details of individual institutional affiliation. We also list a fifth category: the Management Committee. It is comprised of representatives from the donor agencies plus three members from civil society organizations. In addition to overseeing the management of the process, Management Committee members participated in writing, reviewing and deliberating over this text.

Civil Society Organizations (including Indigenous Peoples Organizations): Alejandro Argumedo, Margarita Florez, Glen Hearns, Dan Leskien, Atencio Lopez, Andrew Mushita, Gurdial Singh Nijar, Rene Salazar, Priscilla Settee, Hope Shand.

Private Sector/Industry: Don Duvick, Klaus Leisinger, Brian Lowry, Radha Ranganathan, P. V. Subba Rao, Tim Roberts, Reinhard von Broock.

Public Sector: Tewolde Egziabher Gebre Behran, Lim Engsiang, Geoffrey Hawtin, Gesa Horstkotte-Wesseler, Mita Manek, Nora Olembo, Nuno Pires de Carvalho, Theo van de Sande, Louise Sperling, Ricardo Torres, Vo Tuan Xuan.

Academic: Assiah Bensalah Alaoui, Carlos Correa, Michael Flitner, Cary Fowler, Jaap Hardon, Francisco Martinez-Gomez, Michel Pimbert.

Management Committee (donors plus three CSOs): Susan Bragdon, Chusa Gines, Christine Grieder, Michael Halewood (co-ordinator), Pat Mooney, Olle Nordberg, Vicky Tauli-Corpuz, Carl-Gustaf Thornström, Beate Weiskopf, Joachim Voss (Chair).

We would like to extend a special thank you to Bernard Le Buanec who spent many hours pouring over this text and making so many wise and helpful suggestions. Mr Le Buanec's contributions are very much appreciated by everyone who has been involved in the process. We would like to acknowledge the contribution and valuable advice of Jose Esquinas-Alcazar who attended meetings and participated as an observer. Finally, the Group extends its thanks to Bo Bengtsson for attending two of the group's meetings and for his participation in the process overall.

The entire Group wishes to express its gratitude to Dr Joachim Voss of IDRC who has chaired the Crucible II proceedings. The Group also wishes to thank IDRC for making available Michael Halewood, who served as the brilliant, delightful and indefatigable executive secretary throughout the whole process. Michael Halewood's intellectual and inspirational contribution kept all members working hard through many long nights. The Group also wants to thank Hope Shand of RAFI for her absolutely crucial role in making Volume 1 of these publications a success. It is not exaggerating to say that without Hope Shand, this volume simply would not exist.

Finally, the Crucible Group wishes to thank the following organizations for their financial support and active participation in the meetings and discussions: BMZ/GTZ (German Federal Ministry for Economic Co-operation and Development/German Technical Cooperation), Germany; CIDA (Canadian International Development Agency), Canada; DHF (Dag Hammarskjöld Foundation), Sweden; IDRC (International Development Research Centre), Canada; SDC (Swiss Agency for Development and Cooperation), Switzerland; Sida-SAREC (Swedish International Development Cooperation Agency), Sweden.

List of
Crucible Recommendations

Introduction
The Struggle for Genetic Resources

'Biological diversity' is broadly defined as the diversity of life on earth.[1] Although the abbreviated term 'biodiversity' was scarcely known prior to 1988, today it is ubiquitous — a popular, trendy phrase used by policy-makers, citizens and the mass media worldwide. Popular, but vastly under-appreciated, biodiversity plays a profound role in our daily lives. An estimated 40% of the world's economy is based on biological products and processes. We all depend on biological diversity for our survival, but the Crucible Group is particularly concerned about the loss of biodiversity, especially the biological diversity of farmers' varieties, and its possible impact on food security for the world's poor, who rely on biological products for an estimated 85–90% of their livelihood needs (food, fuel, medicine, shelter, transportation, etc.). For example, over 1.4 billion rural people — primarily poor farmers — depend on farm-saved seeds and local plant breeding/selection as their primary seed source; billions more rely on seeds developed by commercial plant breeders. More than three-quarters of the world's population rely on local health practitioners and traditional medicines for their primary medical needs. At the same time, over half of the world's most frequently prescribed drugs are derived from plants or synthetic copies of plant chemicals.[2]

Though commonly perceived as an environmental issue, biodiversity is profoundly political. Who should be able to own and control the various components of biodiversity (e.g. plants and plant parts at the genetic, varietal, and species levels)? And under what circumstances? How can we best conserve and utilize biodiversity? How will access to genetic resources be regulated, and how will benefits derived from its use be shared equitably and sustainably? Who will decide?

The term 'intellectual property' (IP) refers to a variety of rights granted by a state authority to protect inventors or artists from losing control over their ideas or innovations.[3] Not long ago, the term was scarcely heard outside of a small circle of inventors, government bureaucrats, and patent lawyers. Today, it is widely discussed in the context of international trade, agriculture and development. Does IP play a role in enhancing biodiversity, is it neutral, or will it have a negative impact? Does intellectual property promote innovation and dissemination of knowledge? Is it a tool that can be used to protect the knowledge and biological resources of indigenous and local communities? Should the products and processes of life be patented? Who will decide?

See p. 111 *et seq.* for the Notes

The current challenge

The intersection between intellectual property and biodiversity, giving rise to a policy debate over its wider implications for society, is a subject of conflict and uncertainty at the dawn of the 21st century. The challenge is formidable because science, technology, and social and legal thought that relate to biodiversity and intellectual property are evolving rapidly. Policy-makers are faced with active, complex debates relating to IP and biodiversity in multiple intergovernmental fora. Advances in science and technology have changed the way society uses and values biological diversity. The scope of a growing number of countries' intellectual property laws is being expanded, under prescribed circumstances, to include a variety of biological materials and processes (e.g. those that can be said to be novel, useful and non-obvious if they are inventions; distinct, uniform and stable, and in some cases, discovered, if they are plant varieties). In most cases, these expansions of scope are being undertaken in compliance with minimum standards established in international trade agreements such as TRIPs (Trade-Related Intellectual Property Rights) and NAFTA (the North American Free Trade Agreement). Consequently, policy-makers are being faced with the daunting task of drawing a line in the sand between those biological materials and processes that can be made subject to intellectual property protections, and those that should not.

The debate surrounding control and ownership of intellectual property in biological resources spans local communities, national governments and intergovernmental organizations. At all levels, there are enormously diverse actors and stakeholders, and, not infrequently, intense conflicts between them. There are differences of opinion regarding how the benefits of biodiversity should be shared, and whether or not (and to what extent) biological materials should be subject to intellectually property claims. For some governments, policy-makers, scientists, private sector representatives, and civil society organizations (CSOs),[4] the subject is perceived primarily as a finance and trade issue; for others it is a topic relating to agriculture, food security and human rights; for still others it is debated in the context of the environment and development.

Balancing immediate obligations and long-term commitments

Timing is critical, not only because national governments are faced with immediate legal obligations under international trade and environment treaties, but also because loss of biological diversity, and particularly that of farmers' varieties, is accelerating. Despite heightened appreciation and awareness of biodiversity, and the crafting of international treaties designed to conserve it, the loss continues. We are losing the options we need to strengthen food security and survive global climate change.[5]

Nearly two decades of debate surrounding agricultural biodiversity have firmly established that all nations of the world are genetically interdependent in terms of access to genetic resources. Whether agricultural genetic diversity is found in farmers' fields, in private collections or in high-tech gene banks, no country is self-sufficient in plant or animal genetic resources. Even the most genetically abundant nations of the world look beyond their own borders for at least half of the germplasm required for their staple foods. This reality underscores the need for international cooperation.

In 1996 the world community adopted the Leipzig Global Plan of Action, a blueprint for sustainable management and use of plant genetic resources. The Global Plan has not been fully implemented. Ultimately, a coherent global policy for conserving and utilizing genetic resources is not possible without commitment from the international community.

The Convention on Biological Diversity (CBD) establishes that biological diversity is subject to national sovereignty, but rules governing access to biodiversity and benefit-sharing are still being negotiated. To what extent will there be an internationally agreed regime? Will access be determined by a multilateral system, bilateral agreements, or both?

Encouraging scientific innovation and promoting the public good

The issues of intellectual property and biodiversity are influenced by the larger trends of globalization and privatization. Increasingly, the development and use of knowledge is proprietary. The roles of public and private sector agricultural research have shifted dramatically. The past decade saw dramatic consolidation in the life sciences industry, with market shares of bio-industrial products related to commercial agriculture, food and health tightly concentrated in the hands of just a few transnational enterprises. Should we not strengthen and protect the international public good associated with the optimal flow of germplasm? What are the most appropriate mechanisms for encouraging scientific innovation?

Amid the process of globalization, new rules and new actors are changing governance structures. Multilateral rule-making in a global marketplace is influencing (some would say eroding) the role of the nation state. Will the role of civil society organizations and national sovereignty be restricted or enhanced by these developments? How do we insure a genuinely 'level playing field' for all governments with respect to access to information, and equitable participation in relevant negotiating fora related to biodiversity?

The 1994 agreements between the international gene banks of the Consultative Group on International Agricultural Research (CGIAR, or CG) and the Food and Agriculture Organization of the United Nations (FAO) provide legal recognition that the world's most important collection of genetic resources for food and agriculture is held in trust for the world community. In an era when an increasingly large proportion of global research and development is subject to intellectual property rights, and the CG's research budgets are in decline, can the FAO/CGIAR Trust Agreement protect these genetic resources and insure that they remain in the public domain?

See p. 111 *et seq.* for the Notes

Today, knowledge is perhaps the most important factor determining a nation's standard of living — more than land, tools or labor.[6] In some cases, the growing knowledge gap between North and South is exacerbated by the privatization of the knowledge enterprise. Eighty percent of the world's commercial research and development (R&D) and a similar share of its scientific publications come from the more industrialized nations.[7] World Bank vice-president Ismail Serageldin warns of an 'emerging scientific apartheid'.[8]

The strengthening of intellectual property regimes and their extension to biological materials creates both opportunities and concerns for developing countries. The World Bank's 1999 World Development Report observes that stronger intellectual property rights (IPRs) are a 'permanent feature of the new global economy'.[9] While IPRs are widely accepted as an important tool for stimulating domestic R&D, the report also notes that there is limited empirical evidence that stronger IP regimes lead to increased investments in R&D, even in industrialized countries.[10]

The World Development Report points out that stronger IP regimes, often covering fundamental research tools as well as marketable products, may lead to a higher cost of acquiring knowledge, and could erect barriers to the participation of new firms and researchers in the developing world. There is a concern that stronger IP regimes may actually slow the overall pace of innovation, and increase the knowledge gap between industrial and developing countries.[11] Once again, there is limited empirical evidence to confirm this, just as there is very little on the positive impact of IPRs on increased R&D.[12] Some believe that the knowledge gap can only be narrowed by promoting transfer to developing countries from technology owners in the industrialized world, and that a firm IP framework is an essential precondition for this to happen. 'A desirable IPR regime', concludes the World Development Report, 'is one that balances the concerns of all parties affected by strengthened IPRs'.[13]

Balancing rights and responsibilities

There is growing recognition that the innovation of farmers and indigenous peoples is of utmost importance in understanding, utilizing and conserving biological diversity. This principle is a prominent feature of the CBD and of Farmer's Rights as enshrined in FAO's International Undertaking on Plant Genetic Resources.[14] The Draft Declaration on the Rights of Indigenous Peoples also recognizes the rights of indigenous peoples over their cultural and genetic resources.

The World Trade Organization's Agreement on Trade-Related Intellectual Property Rights (TRIPs) brings intellectual property to center stage in international trade negotiations and obliges member nations to implement national IP laws for plant varieties and other biological materials. However, existing IP regimes do not recognize or protect the rights of informal

See p. 111 *et seq.* for the Notes

innovators over their genetic resources and knowledge. Some people believe that existing IP regimes appropriate the genetic resources and knowledge of farmers and indigenous people. Recent IP claims on plants and human genetic material, for instance, have provoked charges of 'biopiracy' in many regions of the world. Others insist that claims of biopiracy are based on ignorance or a misunderstanding of the principles and application of IPRs. Can national governments meet their obligations to international trade agreements while fulfilling their responsibility to recognize, protect and promote the knowledge and resources of farmers and indigenous peoples? Is a regime for the protection of community knowledge compatible with existing IP regimes?

The development of new genetic technologies, stimulated, in part, by intellectual property protection, has led to the commercial introduction of biotechnological products for agriculture and human health. However, international IP laws covering new seed technology increasingly allow for national laws to restrict the right of farmers to save and re-use proprietary seed. Similarly, scientists are developing genetically engineered plants that are designed to yield sterile seeds. Genetic seed sterilization, if commercialized, could potentially restrict the ability of farmers to save and re-use seed from their harvest. Do these laws and technologies threaten biological diversity, global food security and the innovative capacity of farmers? Or will they have a positive impact on biological diversity?

International policy issues
The Crucible Group re-convened in 1998 because of the urgency of wrestling with these complex and controversial issues. In the following pages we offer a summary of the facts, the fights and the fora relevant to genetic resources and intellectual property. We review some of the major sociopolitical developments that influence and inform the biopolicy debate, as well as the technologies that are transforming our ability to decipher, use and engineer the genomes of all living organisms. Ultimately, the objectives of this report are to distill information for decision-makers, to promote a level playing field for all, and to move the policy debate forward.

The Crucible Group deliberated long and hard about how to present most effectively the interrelated topics covered in this volume. Should major scientific developments be introduced before legal and policy trends were considered, or vice versa? In reality, scientific developments have very direct political implications. Conversely, many scientific and technological trends are spurred by institutional or political developments. In the end, one could easily argue for a different ordering of the chapters, or a different presentation of the topics within chapters. As this document illustrates, science and politics are inextricably linked, more so than ever before.

See p. 111 *et seq.* for the Notes

Part One: Context

A Wider Lens for Considering Biodiversity and Intellectual Property Issues

Introduction

As we enter the new millennium, globalization and privatization are among the most obvious and fundamental trends that affect and influence policy debates on ownership, conservation and exchange of biological materials. The past two decades have witnessed increasing privatization of agricultural research and development, expansion of the scope of intellectual property rights to cover biological products and processes, and liberalization of global markets. These trends have stimulated the commercial development of biotechnology products for agriculture and human health, and the concentration of economic power among a handful of giant life sciences corporations. The process of globalization influences not just the economy, but also culture, technology and governance relating to biological materials.

Globalization and trade liberalization are fueling economic growth, increased prosperity and new opportunities. World exports of goods and services nearly tripled between the 1970s and 1997. Foreign direct investment exceeded (US)$400 billion in 1997, seven times the level in the 1970s.[15] However, there is growing disparity between poverty and privilege, both within and between countries and regions. Thus far, the benefits of globalization are uneven. The United Nations Development Programme (UNDP) concludes that the top fifth of the world's people in the richest countries accounts for 82% of the expanding export trade and 68% of foreign direct investment. The bottom fifth of humanity in the poorest countries account for only 1%.[16] While millions of people are integrated and empowered by knowledge and communication technologies such as the World Wide Web, others remain isolated and marginalized.

New actors, new roles and new rules are redefining global governance. The World Trade Organization (WTO), for example, and the multilateral agreement on intellectual property it administers are reducing the scope for national policy in that arena. Multilateral rule-making in a global marketplace is changing the role of state sovereignty. There is concern that countries and local communities will be increasingly restricted in their ability to determine domestic standards for regulatory and environmental protection, for instance.

The combined pressures of poverty, population growth and environmental degradation pose daunting challenges for agriculture and human

See p. 111 *et seq.* for the Notes

development, especially in the developing world. Over 800 million people in the world today are chronically undernourished.[17] An estimated 1.3 billion people live on incomes of less than one dollar a day.[18] Within ten years, more than half of the world's population will be living in cities.[19] By the year 2020 there will be an additional 2 billion people to feed. Members of the Crucible Group may not agree on the underlying causes of poverty and hunger, but it is clear that different visions of agricultural development are emerging to meet the challenge of global food security and sustainability.[20]

Given limited possibilities for expanding cultivated land area, the Crucible Group agrees that our future food security depends upon a combination of carefully crafted production and distribution policies combined with scientific strategies that ally farmer-researchers with formal sector plant breeders and laboratory-researchers to maximize germplasm enhancement and farming systems. Beyond this, however, policy and program choices often differ. The conventional approach to closing the food gap emphasizes the role of high-input, industrial-scale farming, perhaps complimented by commercial biotechnologies to raise yield ceilings. A second perspective, sometimes described as the 'Double-Green Revolution' approach, proposes sustainable crop production with fewer chemicals and with plant varieties designed to increase resistance to insects and diseases, and with drought tolerance and improved nutritional qualities. Others consider the 'Double-Green Revolution' to be little more than 'business as usual' for the multinational chemicals industry and argue instead for biodiversity-based agricultural research emphasizing local and regional self-sufficiency in food production, focusing primarily on the needs of limited-resource farmers in marginal farming environments. This approach stresses the contribution of farmer-led initiatives, the use of crop varieties developed by farmer/breeders in partnership with formal sector plant breeders, and the use of technologies that will lessen farmers' dependence on purchased inputs. Not without its critics, this approach is sometimes attacked for its perceived Malthusian naivety or absurd political correctness. Still others would see policy as by far the most important engine of change and argue that sound, 'pro-poor' choices in land tenure, credit and price subsidization are the key to food security.

No matter what combination of germplasm, technologies, farming systems and policies are employed to achieve food security in the 21st century, commercial biotechnology and subsistence farmers alike will need to make better use of a broader range of the world's plant and animal genetic diversity. Farmers will require crop varieties and livestock breeds capable of producing under diverse and rapidly changing conditions. Under any scenario, genetic resources for food and agriculture, the role of farming communities who nurture and develop diversity, and the vital contributions of formal sector plant breeders, all assume critical importance.

This chapter introduces some of the major social, economic and

See p. 111 et seq. for the Notes

environmental trends that influence or inform the larger debate on ownership, conservation and exchange of biological materials. What are these trends that influence the way society thinks about, values and uses biological diversity at the beginning of a new millennium?

The accelerating loss of biological diversity

There is increasing awareness worldwide of the value, importance and fragility of biological diversity. In the final decade of the 20th century, greater numbers of people became aware of species extinction, erosion of genetic resources and the threat of ecosystem destruction. The concept of biodiversity has entered the mainstream of government thinking and, in some cases, traveled beyond national ministries of environment. Despite heightened appreciation and awareness of biodiversity and the crafting of international conventions designed to conserve it, the loss of biological diversity continues. Forests are falling, fisheries are collapsing, plant and animal genetic diversity is eroding all over the world.

- Loss of diversity in plant genetic resources for food and agriculture has been substantial, and these resources are disappearing at unprecedented rates. No one knows how much diversity once existed in domesticated species, so it is impossible to say exactly how much has been lost historically.[21]

- Domestic animal breeds are disappearing at an annual rate of 5%, or six breeds per month.[22] According to FAO, the status of almost one-third of all livestock breeds is endangered or critical.

- Tropical forests are falling at a rate of just under 1% per annum, or 29 hectares per minute.[23] From 1980 to 1990, the loss was equivalent to an area the size of Ecuador and Peru combined.

- All of the world's main fishing grounds are being fished at or beyond their limits. About 70% of the world's conventional marine species are fully exploited, overexploited, depleted or in the process of recovering from overfishing.[24] During the 20th century, about 980 fish species became threatened.[25]

- Nearly 60% of the earth's coral reefs are threatened by human activity.[26]

- A new study conservatively estimates that 34 000 species of plants — 12.5% of the world's flora — are facing extinction. At least one of every eight known plant species on earth is threatened.[27]

- For every plant that becomes extinct, 30 other species go with it — many of them are microorganisms. Some biologists warn that plant pathogens (including fungi, viruses and bacteria) should be conserved with the same urgency as other species because they play a vital role in the functioning of ecosystems.[28]

The loss of biodiversity threatens food security, especially for the poor, who rely on biological products for 85–90% of their livelihood needs (i.e. food, medicine, fuel, fibre, clothing, shelter, energy, transportation, etc.).

See p. 111 *et seq.* for the Notes

New estimates from FAO indicate that there are 828 million chronically undernourished people in the world, a small increase since the early 1990s.[29]

Erosion of cultural diversity

The 1993 Crucible Group concluded that we cannot conserve the world's biological diversity unless we also nurture the human diversity that protects and develops it. Today, there is growing recognition that loss of *cultural diversity* — of traditional farm communities, languages and indigenous cultures — is intricately linked to the loss of biological diversity. Many members of the Crucible II Group are alarmed by the loss of the culturally-based knowledge represented by thousands of diverse cultures that are themselves endangered or disappearing.

Globally, language is one of the strongest indicators of cultural diversity. The highest levels of plant and animal diversity, as well as the world's richest linguistic life, are found close to the equator. Ten out of 12 'megadiversity' countries identified by the International Union for Conservation of Nature and Natural Resources (IUCN) rank among the top 25 countries for 'endemic' languages (i.e. languages spoken exclusively within a country's borders — which usually means the majority of the smaller languages of the world).[30] Some cultures have many distinct names describing a single plant, its parts and its uses. The diversity of names associated with distinct properties of a species is multiplied by the number of languages and dialects spoken by distinct communities that use the same biological resource.[31] With the loss of their language, the community loses its ability to describe and therefore to use the plant. While loss of knowledge does not imply the loss of the plant itself, this is commonly the case, since with the decline in the knowledge of the uses of the plant, the community may lose interest in its conservation.

As with biological diversity, the magnitude and pace of the current 'extinction crisis' in linguistic diversity is unprecedented. Linguists who monitor the status of surviving languages conclude that approximately 6–11% of the 6703 languages spoken in the world today are 'nearly extinct', and they predict that 50–90% will disappear during the 21st century.[32]

Local and indigenous peoples who speak ancestral languages are severely threatened by loss of sovereignty over land, resources and cultural traditions, and the promotion of linguistic assimilation. As they become increasingly marginalized, local people lose local scientific knowledge, innovative capacity, and wisdom about species and ecosystem management.[33] As one scholar concludes: 'Any reduction of language diversity diminishes the adaptational strength of our species because it lowers the pool of knowledge from which we can draw.'[34]

The loss of traditional farm communities, languages and indigenous cultures all represent the erosion of human intellectual capital on a massive scale. It is tantamount to losing a road map for survival, the key to food

See p. 111 *et seq.* for the Notes

security, environmental stability and improving the human condition. Thus, it is increasingly difficult to talk about the conservation and sustainable use of genes, species and ecosystems separate from human cultures.

On-farm conservation and use of plant genetic resources

Although many in the scientific community are used to measuring progress through precisely defined achievements, one of the genuinely significant 'breakthroughs' of the past five years owes more to rediscovery (by some) than discovery. Since 1993, agricultural research, including plant breeding and germplasm conservation, has been revitalized by the creative association of farmer-innovators, and their communities, with formal sector scientists and research institutions. The management and enhancement of plant genetic resources has been in the hands of farmers from the beginning of agriculture. Fortunately, there is far greater recognition today than there was five years ago that the contributions of farmers and indigenous peoples are critical to the conservation, use and active enhancement of biological diversity, and that these groups and individuals should be recognized and rewarded for their contributions. This principle is a prominent feature of the Convention on Biological Diversity (CBD) and of Farmers' Rights as discussed and supported in the FAO Commission on Genetic Resources for Food and Agriculture (CGRFA). The innovative activity of farmers with respect to the development of crop varieties and farming methods is described variously: research, plant breeding, ethno-science, informal innovation, and so on. Not everyone agrees on which is the best term. As a collective, the Crucible Group wants to avoid getting bogged-down in terminological wrangling. In the end, what is important is that the Group recognizes the value of farmers' innovations.

In the field of agricultural genetic resources, there is greater appreciation for the fact that *in situ* (on farm) conservation is a crucial element in the conservation of agricultural biodiversity and that it must be complementary to gene bank collections. When *ex situ* germplasm is removed from its cultural and environmental context, it loses the ability to adapt to constantly evolving pests, diseases and the ever-changing needs of local communities. By placing greater emphasis on *in situ* and farmer/community level management of genetic resources, both the CBD and the Leipzig Global Plan of Action for Plant Genetic Resources for Food and Agriculture (PGRFA) emphasize that the future of world food security depends not just on stored crop genes, but on the people who use and maintain diversity on a daily basis. Leipzig's Global Plan of Action provides the first intergovernmental recognition and support for on-farm management and improvement of plant genetic resources. It recommends new on-farm conservation and participatory breeding initiatives, including the need for stronger links between conservation and utilization of plant genetic resources.

See p. 111 *et seq.* for the Notes

'Participatory plant breeding' (PPB) is still in its youth, but spreading quickly.[35] PPB is a new approach to germplasm development and conservation involving scientists, farmers and other end-users (i.e. rural cooperatives, consumers, extension workers, etc). It is termed 'participatory' because users have a research role in all major stages of the breeding and selection process. PPB is a crop improvement strategy that especially seeks to involve disadvantaged user groups (i.e. women and other resource-poor farmers). Over 70 cases of participatory plant breeding are documented, involving a range of crops and geographic regions.[36] These include, for example, pearl millet in India, barley in Syria, common beans in Brazil, rice in Nepal and cassava in Colombia.

Support and recognition for on-farm conservation and farmer-driven breeding is growing. A range of strategies is being developed to enhance genetic materials on-farm, for and by farmers. Examples include: CGIAR's Systemwide Program on Participatory Research and Gender Analysis (and its working group on participatory plant breeding),[37] the Community Biodiversity and Development Conservation Program (CBDC), the Seeds of Survival Program in Africa, the Academy of Development Sciences in India and Projecto en Tecnologia Alternativa (PTA) in Brazil. Efforts to conserve and enhance germplasm systems are found for both minor and major crops, and in 'marginal' farming areas (i.e. those with poor soil, little rainfall, steep hillsides) as well as irrigated lands.

The increase in participatory plant breeding and other collaborative programs involving farmers, their communities and formal sector scientists raises new questions and challenges for recognizing collaborative innovation in plant breeding. Some observers believe that neither Farmers' Rights nor Breeders' Rights adequately address these issues. The International Development Research Centre (Canada) has recently funded work to address the property rights, best practice, and ethical dimensions of community-based and formal collaborations.

Many members of the Crucible Group agree that there is a need to strengthen the role of indigenous and local communities in order to ensure their full participation in germplasm conservation and enhancement.

Global climate change and biodiversity

Although there is no consensus among scientists, there is growing international opinion that global climate change will have profound impacts on biodiversity and will compromise the sustainability of human development on the planet. The Intergovernmental Panel on Climate Change (IPCC) predicts that the build-up of greenhouse gases in the atmosphere will cause global temperatures to rise by 1–3.5 degrees Centigrade during the next century; and that melting glaciers and thermal expansion of the ocean will bring an associated rise in sea level of between 15 and 95 centimeters.[38] Simulation models predict that each one-degree rise in

See p. 111 *et seq.* for the Notes

temperature will displace the adaptation of terrestrial species some 125 km towards the poles, or 150 metres in altitude. Approximately 30% of the earth's vegetation could experience a shift as a result of climate change. But since climate will change faster than the migration rate of many species, models predict a 'drastic reduction' in global species diversity.[39]

New research on the impact of global warming on vegetation offers especially grim predictions for the tropics. According to the Institute of Terrestrial Ecology (Edinburgh, UK), by 2050 a warming of up to 8 degrees Centigrade in parts of the tropics will lead to higher evaporation rates, lower rainfall and eventually the collapse of tropical ecosystems. The World Bank estimates that a 2–3 degree rise in global mean temperature will reduce the mass of mountain glaciers by one-third to one-half, and endanger at least one-third of all species surviving in forests.[40] Changes in glacier mass and forest area will have a profound impact on agricultural productivity. Millet crop yields in Africa are expected to drop by between 6% and 8%; a Senegalese study predicts that millet yields in Senegal will decrease by between 11% and 38%.[41] In South Asia, yields for rice and wheat are expected to fluctuate wildly. The maize crop in South Asia and Latin America may shrink by between 10% and 65%.[42]

Not all scientists (and not all members of the Crucible Group) agree with grim predictions for global climate change, and some point out that substitute crops may be developed to offset shrinking yields and increase productivity in some regions. Some models have shown that global warming could be neutral or favorable in temperate latitudes and disadvantageous for the tropics and sub-tropics. The discrepancies and uncertainties in climate models do not permit more accurate predictions. Regional projections suggest that Africa might be the continent worst hit by climate change.[43]

Ultimately, biological diversity is the key to adapting to global climate change. If we are to adapt food production systems to radically changing conditions in the coming decades, plant and animal diversity will be the single most critical resource for doing so.

The changing roles of public and private sector agricultural research

Until recently, agricultural research has been largely publicly financed and its products made freely available. Today, agricultural research is increasingly privatized. A rapidly changing IP environment and declining research budgets have marginalized the role of public sector agricultural research in both industrialized and developing countries. We are living in an era when, increasingly, science is subject to property rights and knowledge is commodified.

Given the demonstrated success of agricultural research in bringing science and technology-based solutions to agricultural production constraints, it is paradoxical that public agricultural research is now facing a global crisis. Many national agricultural research institutions in developing countries suffer from lack of money, even to buy the basic essentials or to pay salaries. Institutions in industrialized countries and the international research centres of the Consultative Group on International Agricultural Research (CGIAR) face similar financial constraints.[44]

The crisis is not only financial, it is also one of confidence. In spite of the demonstrated effectiveness of agricultural research in the past, policy-makers today are not convinced that further investments are necessary or that they can accomplish the seemingly overwhelming tasks of assuring food security to millions of the poorest of the poor. The end result is that financial allocations to agricultural research are reduced at a time when its importance to economic development remains critical. Reduced support leads to low productivity, low visibility and, ultimately, less political support for the agricultural research system — a vicious circle.[45]

Public funds for agricultural research have stagnated or declined, while private investments have increased at an unprecedented rate:

- In the OECD, for example, private agricultural R&D totaled $7 billion in 1993 compared with $4 billion in 1981, an annual rate of growth of 5.1%. By contrast, publicly performed agricultural R&D rose by 1.7% per annum, from $5.7 billion in 1981 to $6.9 billion in 1991.[46]
- The relative importance of private R&D in total agricultural R&D varies across the OECD countries, but exceeds 50% in seven countries, including the US and Japan, which together account for over one-half of all private agricultural research throughout the OECD. The private share in the remaining OECD countries is smaller (about one-third in 1993), but has been growing rapidly. Private sector R&D in developing countries typically accounts for 10–15% of total agricultural R&D.[47]
- Private and public agencies perform different types of R&D. About 12% of private agricultural research focuses on farm-level technologies, while 80% of public research has that orientation. Food processing and post-harvest research is the dominant focus of private agricultural research, accounting for 30–90% of all private agricultural R&D.[48]

The increased role of the private sector is propelled, in part, by dramatic advances in biotechnology and the strengthening of intellectual property protection for biological materials in many countries. Until about a decade ago, market failure in agricultural R&D seems to have been widely taken for granted. The main reason was inappropriability of benefits.[49] Constraints to private investment in agricultural research have changed dramatically: with advances in biotechnology, the process of technology generation has accelerated, making it possible to reap the benefits of investments relatively soon. Since 1980, the evolution of intellectual property laws has allowed for the patenting of living organisms, enabling biotechnology companies to patent biological processes and products,[50] thereby increasing incentives for the private sector to invest in this type of research.

See p. 111 *et seq.* for the Notes

With globalization of trade and the disappearance of trade barriers, the market for biotechnology products has expanded, providing increased opportunities for large multinational corporations. The role of multinational corporations has both positive and negative implications for global science. From a positive point of view, these firms have strong research and development arms that are involved in both applied and basic research, frequently involving scientists from many countries and benefits that extend across borders. However, the products of private sector research are generally proprietary and may not be accessible to the poor or those who need them.[51]

The decline in public sector agricultural research budgets is prompting new partnerships between the public and private sectors, especially in the field of agricultural biotechnology. For example, Swiss agrochemical giant, Novartis, and the University of California at Berkeley signed a $25 million, five-year agreement in November 1998. Although the agreement specifies that Novartis cannot dictate what research will be performed with its money, the company will have first rights to negotiate an exclusive license on 30–40% of any inventions made in the Department of Plant and Microbial Biology.[52] Private–public interactions can create fertile ground for knowledge generation and transfer of technology to the marketplace. But private financing of the public sector is not without controversy. Critics charge that such alliances will give private companies the ability to influence the research agenda at publicly funded institutions, allowing public goods to be appropriated for private profit.

The CGIAR's agricultural research centres are also discussing new models of collaboration with the private sector (joint research projects, strategic alliances, etc.) and policies relating to the use of proprietary materials and technologies. For example, a B.t. (*Bacillus thuringiensis*) gene for insect resistance provided by Plant Genetic Systems (Belgium) has been transferred to potato varieties, and the International Potato Centre has permission to distribute them in ten developing countries. Monsanto has entered into agreements with Kenyan and Mexican agricultural research institutes to develop virus-resistant crops. While these collaborations are successful, they are few in number, highly bilateral and often components of philanthropic programs.[53]

The potential neglect of the public good is one of the primary issues raised by the dramatically changing roles of the public and private sector in agricultural research. There is concern that the formerly open exchange of materials and technologies to help the poor is being constrained and complicated by intellectual property. Over the past five years, the New York-based Rockefeller Foundation has invested $50 million in plant biotechnology for the developing world. In the words of Dr Gordon Conway, President of the Rockefeller Foundation:

"As plant research in the industrialized world has come to be dominated by private companies who closely guard their proprietary technologies, the process of innovation in the developing countries has slowed. Public sector plant breeders don't know how to respond, and when they try, they are handicapped by the huge disparity in resources and negotiating power between themselves and the companies."[54]

Crucible Recommendation 1

Broadening research participation
The Crucible Group believes that it is both possible and necessary to increase participation and partnership in agricultural research and in genetic resource conservation. The Group recommends that:

- governments increase their contributions and strengthen their long-term commitments to national and international agricultural research and genetic resource conservation;
- governments and non-governmental organizations increase financial support and institutional commitment to all agricultural research that involves participatory plant breeding and to on-farm genetic resource conservation;
- all concerned parties give support to innovative initiatives to encourage effective and equitable forms of partnership collaboration in agricultural research.

Crucible Recommendation 2

Support for in situ *or on-farm conservation*
We recognize the complimentarity of *ex situ* and *in situ* conservation strategies. The two have very different effects in terms of which germplasm endures and how that material evolves, yet both are essential to safeguard germplasm for future generations and both require increased support. In particular, *in situ* or on-farm conservation deserves increased support as few approaches have yet been identified that are financially, socially and technically efficient, effective and acceptable to those who primarily undertake *in situ* conservation, that is, farming communities themselves. Increased support for *in situ* or on-farm conservation activities need not displace or diminish funding for *ex situ* conservation.

Crucible Recommendation 3

Support to farmer-led science and participatory plant breeding
Organizations and governments should formulate policies and actions to develop and promote more farmer-based and farmer-led science, including participatory plant breeding. Sustainable agriculture, including dynamic crop management, can better be achieved if farmers' own creative capacities are stimulated and strengthened (financially and technically) to partner with the more formal research and development efforts.

See p. 111 *et seq.* for the Notes

How can the benefits of technology that is subject to intellectual property rights be harnessed for the needs of the poor and the environment? How can public systems, both nationally and internationally, insure that research priorities are not unduly influenced by private companies, and that international public goods remain in the public domain? Of course, there are no guarantees that the public sector is benign or always acts in the public interest. Conversely, it is unfair to assume that the private sector is acting against the public good.

Consolidation in the life sciences industry

Consolidation is taking place in all sectors of the global economy. In 1997, the value of all mergers and acquisitions hit a staggering $1.6 trillion in worldwide deals, up from $454 billion in global activity recorded in 1990.[55] In 1998, the total volume of mergers and acquisitions worldwide was a record $2.4 trillion — a 50% increase over 1997.[56] According to the UN Conference on Trade and Development, over four-fifths of all foreign direct investment worldwide is now in the form of global mergers and acquisitions.[57] The past decade recorded dramatic consolidation in the 'life sciences', with market shares of bioindustrial products related to agriculture, food and health tightly concentrated in the hands of giant transnational enterprises. For example:

- the world's top ten agrochemical corporations account for 91% of the $31 billion agrochemical market worldwide;[58]
- the top ten global seed companies control an estimated one-quarter to one-third of the $30 billion commercial seed trade;[59]
- the top ten pharmaceutical companies account for 36% of the $251 billion global pharmaceutical market;[60]
- the top ten firms hold 61% of the animal health industry market valued at $16 billion.[61]

Under the life sciences banner, many firms are using complementary technologies to become significant actors in all of these categories.[62] Traditional boundaries between pharmaceutical, biotechnology, agribusiness, food, chemicals cosmetics, and energy sectors are blurring and eroding. Major transnational companies are restructuring to take advantage of the molecular revolution and the complementary use of technologies such as high-throughput screening, combinatorial chemistry, transgenics, bioinformatics and genomics. Life sciences companies are securing and protecting information and technology via patents, and, in some cases, that quest is driving a restructuring of the industry. In today's knowledge-based economy, intellectual property assets have surpassed physical assets such as land, machinery or labor as the basis of corporate value.[63] At the end of 1995, for example, the Hoechst group held 86 000 patents and patent applications.[64] According to Dr Richard Helmut Rupp, head of Hoechst R

See p. 111 *et seq.* for the Notes

& D, 'The most important publications for our researchers are not chemistry journals, but patent office journals around the world.' Demonstrating the value of intellectual property assets, the cover of Novartis' 1997 annual report announces that the company holds more than 40 000 patents.[65] Worldwide demand for patents shows a strong upward trend. At the end of 1995, approximately 3.84 million patents were in force worldwide. The patent offices of the US and Japan together with the European Patent Convention accounted for 81% of the total.

Seed industry concentration

The Crucible Group believes that it is especially important to monitor concentration in the global seed industry, a trend that has been accompanied by dramatic declines in public sector plant breeding. In some industry sectors, particularly in countries of the Organization for Economic Cooperation and Development (OECD), there is evidence of highly concentrated markets. For example:

- the top five vegetable seed companies control 75% of the global vegetable seed market;[66]
- four companies control 69% of the North American maize seed market;[67]
- at the end of 1998, a single company controlled 71% of the US cotton seed market.[68]

Private sector plant breeding and seed sales have been a highly effective tool in many parts of the world to transfer innovation in agriculture, especially through the provision of reliable, clean planting material. Strategies such as market segmentation could play a role in increasing the availability of new crop technologies to poor farmers in the developing world.[69] In the future, however, if access to biotechnological and other plant breeding-related innovations are restricted to a handful of seed companies, the possibility exists for market dominance by a few suppliers, with potentially serious implications for technology choice and price fixing. Free and fair competition may not be possible in the absence of government oversight and regulation, including the use of anti-trust legislation. The option of government anti-trust laws is one mechanism that could be invoked to curb excessive consolidation in the seed industry. Substantial technical assistance would have to be provided to developing countries in order for them to sort meaningfully through the complexities of, and implement, anti-trust laws.

Crucible Recommendation 4

Anti-trust legislation for seed industry

The Crucible Group recommends that anti-trust legislation at the international level be provided and enforced to ensure fair competition practices in the seed industry.

See p. 111 *et seq.* for the Notes

Transgenic crops commercialized

When the first Crucible Group concluded its negotiations in October 1993 genetically engineered crops were not yet sold commercially. Although opinions differ on the ethics and safety of transgenic crops, the commercial market for genetically engineered seeds has expanded dramatically in scale and geographic scope in recent years.

- From 1986–1997, approximately 25 000 transgenic crop field trials were conducted by 45 countries on more than 60 crops and ten traits. Of this total, 15 000 field trials were conducted during the first ten-year period, and 10 000 in the last two years.[70]

- Soybean, maize, cotton, canola/rapeseed and potato were the five principal transgenic crops grown in 1998. Transgenic soybean and maize accounted for 52% and 30% of global transgenic area respectively. Herbicide tolerance was the dominant trait, accounting for 71% of all transgenic crops; insect resistance accounted for 28% of the global transgenic area.[71]

- According to the International Seed Trade Federation, the world market for genetically engineered seeds is expected to reach $2 billion by the year 2000 and will triple to $6 billion by 2005. The Federation predicts that the market for bioengineered seeds will reach $20 billion in the year 2010.[72]

Clive James of the International Service for the Acquisition of Agri-biotech Applications (ISAAA) produced figures showing a sharp rise between 1997 and 1998 in the size of area of planted with transgenic crops in eight countries (see Table 1) and forecast that by the end of 1999, an estimated 40 million hectares would be planted in genetically modified crops worldwide.[74] Proponents of genetic engineering point out that after thousands of field tests and commercial-scale plantings on numerous continents, no major ecological problems have been identified with genetically modified (GM) crops, nor hazards associated with GM foods currently found on the shelf. However, there is concern about possible ecological impacts of transgenic crops, including the possibilities of gene transfer to related species and resistance to biopesticides.

In recent years the 'precautionary principle' has gained prominence in environmental protection and regulatory debates, especially relating to the approval and commercialization of biotechnology products. In very general terms, the precautionary principle says that government regulators have a responsibility to take preventive action to avoid harm before scientific certainty has been established. However, the precautionary principle has no one internationally recognized definition and its status under international law is widely

Table 1: Area of transgenic crops planted (million ha)[73]

Country	1997	1998
USA	8.1	20.5
Argentina	1.4	4.3
Canada	1.3	2.8
Australia	0.1	0.1
Mexico	0.1	0.1
Spain	0.0	0.1
France	0.0	0.1
South Africa	0.0	0.1
TOTAL	11.0	27.8

Source: C. James, ISAAA

See p. 111 *et seq.* for the Notes

debated.[75] Many observers view it as a significant innovation in the arena of risk assessment and environmental protection, although its definition, application, and scope are still evolving.

In February 1999, delegates from 175 countries met in Cartagena to conclude four year's worth of negotiations on a draft international protocol on biosafety, under the auspices of the Convention on Biological Diversity. Given the wide disparity in negotiating positions, it proved impossible to reach consensus on criteria to govern transboundary movement, handling and use of living modified organisms (LMOs), also referred to as genetically modified organisms (GMOs).

Failure to reach agreement in Cartagena, and at follow-up informal biosafety consultations in Vienna in September 1999, has done little to ease concerns of some governments, farmers and consumers over trade in genetically engineered crops and issues surrounding the potential impacts of genetically modified organisms on health, safety and the environment. 1998 and 1999 witnessed unprecedented public debate over the introduction and use of genetically modified products for food and agriculture — especially, but not only, in Europe. Consumer and farmer resistance to genetically engineered products for agriculture is influencing policy-makers, as well as major food retailers and processors; questions surrounding socioeconomic, health and environmental impacts are not resolved. Consumer demand for mandatory labeling of GMO products and more rigorous biosafety regulations are growing in many parts of the world.

- In April 1999, the seven largest grocery chains in six European countries made a public commitment to sell products that contain no genetically modified foods.
- European-based Unilever and Nestlé, two of the world's largest food and beverage corporations, announced that they would reject genetically modified ingredients for their European products.
- US-based Archer Daniels Midland and A.E. Staley Manufacturing Co., two of the world's largest maize processors, announced that they would not purchase any genetically modified maize that is not accepted in Europe.
- In January 1999, Canada's national health agency declined to approve the use of genetically engineered bovine growth hormone, citing concerns over the drug's impact on animal health and welfare.
- A laboratory study by Cornell University scientists published in May 1999 shows that pollen from genetically engineered insect resistant B.t. (*Bacillus thuringiensis*) maize is toxic to caterpillars of monarch butterflies. Dr John Loose, one of the Cornell researchers who conducted the study, cautions: 'Our study was conducted in the laboratory and, while it raises an important issue, it would be inappropriate to draw any conclusion about the risk to Monarch populations solely on these initial results.'[76]

See p. 111 *et seq.* for the Notes

- Following the release of the Cornell study, the Japanese Ministry of Agriculture, Forestry and Fisheries (MAFF) announced that it would suspend approval of B.t. crops in Japan until revised safety protocols are developed for GM crops.[77]

- The Brazilian State of Rio Grande do Sul has declared itself a GM-free state. Legislation is being drafted to ban commercialization and importation of genetically modified seeds. In June 1999 a Brazilian federal court barred the planting and distribution of genetically modified soybeans in Brazil until environmental impact assessments are prepared. Brazil is the world's second largest exporter of soybeans.

- In June 1999, European Union environment ministers stressed the need to implement a more transparent and strict framework concerning risk assessment, monitoring and labeling of GMOs. Ministers from France, Italy, Greece, Denmark and Luxembourg declared that, pending the adoption of stricter regulations, they would take steps to suspend any new authorizations for the growing or marketing of GMOs.[78]

There is a concern that the controversy and debate surrounding the introduction and use of genetically modified products for food and agriculture could jeopardize the future development and release of bioengineered crops that aim to help the poor and malnourished.[79] For example, the Rockefeller Foundation and the European Union are funding the development of genetically engineered rice strains that they believe will combat widespread nutritional disorders (vitamin A and iron deficiencies) afflicting billions of people worldwide.[80] It is understood that, once perfected, the rice strains would be made freely available to agricultural research centres worldwide.

Restrictions on the right of farmers to save seed

National and international institutions, both public and private, are implementing, developing and promoting a variety of legal and technological tools that are designed to give the seed industry greater control and protection over plant genetics and restrict or eliminate the right of farmers to save and re-use seed from their harvest. For example, the 1991 Act of the Union for the Protection of New Varieties of Plants (UPOV) does not mandate an exemption allowing farmers to use farm-saved seed freely as further planting material. If successfully commercialized and widely adapted, genetic seed sterilization technologies may also restrict the ability of farmers to save seed. In some industrialized countries, the commercialization of patent-protected, genetically engineered seeds is altering the relationship of the seed industry to its customer — the farmer — and is changing traditional farming practices. In order to protect its investment and recoup research costs, the seed industry asserts that it is illegal for farmers to save patented seed for replanting. In the United States it is becoming increasingly common for seed companies to require their customers to sign a licensing agreement that prohibits farmers from saving, selling, or reusing patented seed for any purpose — even on their own land. Some companies are aggressively enforcing patent rights on transgenic (genetically engineered) seed technology.[81] Virtually all transgenic seeds are protected by patents. In the United States, utility patents do not provide for the 'farmers' privilege'.

In the developing world, where the majority of farmers depend on farm-saved seed as their primary seed source, the notion of legal or biological prohibitions on seed saving is perceived by some as both alien and life-threatening. Others believe that restrictions on seed saving will act as an incentive for the private sector to invest in developing improved varieties, stimulate plant breeding in the developing world and thereby contribute to food security.

Viewpoint box: Merits and myths of seed saving

Formal seed sector promotes food security
Seed saving denies farmers the opportunity to grow the best and most recent scientific innovations. Were all farmers to abandon this practice, the rate of innovation would grow substantially and costs, relative to benefits, would drop. If anything, agricultural diversity would increase as the range of innovative breeding increases. While the diversity 'in the field' might appear less, the actual diversity 'over time' would increase. Society, through its governments, should unfetter breeders and halt the biopiracy of high tech seed through 'seed saving'.

Seed saving for subsistence farmers
There is an understandable and genuine international debate regarding the merits and myths of seed saving. However, small-scale subsistence farmers should not be obliged to change their age-old practices through legislation or regulation. Any farmer or farming community that traditionally saves less than 20% of the harvest for future sowing or exchange — including trade in the market — should be able to continue this practice without constraint regardless of the source of the seeds involved.

Farmer-based food security
Farmers — notably women farmers — are plant breeders. Farmers exchange germplasm to improve their plant breeding, and this is the basis of local food security. Local breeding creates and conserves diversity important to the world. Poor farmers, especially, breed for environments and needs that standard commercial breeders neither know nor care about. The Farmers' Right to 'save seed' is also associated with the Right to Food and must be fully secured by national governments and the intergovernmental community.

Biopiracy: fact or fiction?
Some people are concerned that the expanded scope of IPRs and their extension to biological materials enables institutions or researchers effectively to appropriate the resources and knowledge of farmers and indigenous communities, especially in the developing world. In recent years, IP claims relating to plants and human genetic material have provoked charges of 'biopiracy' in many regions. Members of the Crucible Group disagree on whether or not, and to what degree, biopiracy is a significant problem. The following viewpoint boxes on biopiracy, and the specific case of basmati rice, illustrate widely divergent views on the subject.

See p. 111 *et seq.* for the Notes

Viewpoint box: What is biopiracy?

Legalistic view

Biopiracy refers to the appropriation of biological resources without the prior informed consent of the local people and/or of the competent authority of the respective state, for access and benefit sharing, under mutually agreed terms. With the implementation of national and international laws governing access to genetic resources and the development of *sui generis* IP laws for indigenous and local knowledge, biopiracy is becoming easier to identify in legal terms.

When properly applied and enforced, IP will actually promote the objectives of the CBD by creating sustainable uses for biomaterials, providing the means for recovering value that can be fairly shared, and promoting technology transfer. This will not hinder traditional uses of biomaterials by indigenous and local communities. While patents are not benefit-sharing mechanisms, they can generate benefits that can be shared with indigenous and local communities through bioprospecting agreements, for example.

Critical view

The appropriation of genetic resources from Third World countries by private, often multinational companies and/or public institutions (or their intermediaries) from industrialized countries is a structural problem which reflects larger questions of equity — both historic and present-day. Biopiracy is not only a legal question, it is primarily a moral question. Even in those cases where companies or institutions follow legally binding rules on access and benefit sharing or sign bioprospecting agreements — it is still biopiracy, and that is because existing legal frameworks are inadequate to protect the rights of farmers and indigenous peoples. Patents and plant breeders' rights are not benefit-sharing agreements.

No plant breeder or genetic engineer starts from scratch when they develop a new plant variety. They are building on the accumulated success of generations of farmers and indigenous people. Biotech companies claim that they 'invented' their genetically engineered plants or new pharmaceuticals. In reality, they are fine-tuning and modifying plants that were developed by anonymous farmers and improved by the more recent contributions of institutional breeders. To claim exclusive monopoly control of these plants (or genes, or traits) is unjust and immoral.

Industry view

Biopiracy is a highly emotive term. Knowledge and materials in the public domain may be freely used by anyone to make further advances: and such advances may properly be protected by IPRs, but only for a limited time. In those rare cases where it turns out that IP claims are based on indigenous knowledge or germplasm, such claims can be challenged and revoked — further evidence that the IP system is working effectively. Without strong IPRs, the world as a whole loses the wider dissemination of a useful technique, because no one will risk the necessary investment in the absence of IP protection.

Real biopiracy is a serious and readily identifiable problem; it refers to the unauthorized use, multiplication or copying of privately owned innovations that are protected by patent or plant breeders' rights. When farmers reuse patented seed without permission or payment of royalties, for example, that is piracy. To insure a level playing field, we need aggressive enforcement (and compliance) of the TRIPs agreement in all countries.

See p. 111 *et seq.* for the Notes

Viewpoint box: The basmati rice patent: biopiracy or invention?

In September 1997, RiceTec Inc., a Texas-based company, was issued US patent no. 5 663 484, entitled 'Basmati rice lines and grains'. Basmati rice has been grown in the Punjab region of India and Pakistan for centuries. Farmers in this region have selected and maintained basmati rice varieties that are recognized worldwide for their fragrant aroma and distinct taste. Right or wrong, the basmati patent has launched a firestorm of controversy.

Classic biopiracy
RiceTec is capitalizing on the genius of South Asian farmers; germplasm is being pirated, as well as the basmati name. RiceTec's US patent applies to breeding crosses involving 22 basmati varieties from Pakistan and India. The patent claims the invention of 'novel rice lines with plants that are semi-dwarf in stature, substantially photoperiod-insensitive and high-yielding, and that produce rice grains having characteristics similar or superior to those of good quality basmati rice grains produced in India and Pakistan.' The sweeping scope of the patent extends to such varieties grown anywhere in the western hemisphere! Specifically, the patent applies to breeding crosses involving 22 farmer-bred basmati varieties from Pakistan and India. These varieties were initially collected in the Indian subcontinent and deposited (among other places) in a US genebank. Not only does the patent claim genetic material that was developed by South Asian farmers, it also usurps the 'basmati' name — which is geographically specific to varieties grown in parts of India and Pakistan, just as 'champagne' is unique to France. Exports of basmati rice (worth $800 million per annum in India alone) could be threatened if they are forced to compete with 'counterfeit' basmati. The patent therefore jeopardizes the livelihoods of thousands of Indian and Pakistani farmers who grow basmati for export.

No biopiracy
The whole flap is based on a misunderstanding. RiceTec's US patent protects the company's seeds and breeding methods in the US alone, it does not patent or trademark the name 'basmati'. The company has no claims on basmati rice anywhere in Asia. There is a misconception that RiceTec's patent would prevent Indian farmers from exporting their product. This is not true. Basmati is a generic term. Just as durum refers to a class of wheat, basmati refers to a class of rice. Even if it were not, no country is obliged to protect it since neither Pakistan or India has legislation protecting geographical indications under the WTO Agreement on Trade-Related Intellectual Property Rights (TRIPs), Article 24.9.

The germplasm used for breeding RiceTec's basmati rice came partly from publicly-operated gene banks in the US. The specific lines are identified in the patent and they are available to anyone for breeding purposes. The germplasm did not come from India; the basmati varieties claimed in this patent were developed using classical breeding over a period of 10 years. Even if the germplasm originated from India, the company simply used the varieties to create a novel product. This is not biopiracy; this is clearly an invention under US patent law! RiceTec's basmati varieties are truly novel; for the first time it's possible to cultivate high-quality, high-yielding basmati in the western hemisphere.

See p. 111 *et seq.* for the Notes

Human biodiversity

The Crucible Group notes that many of the issues now being hotly debated over plant genetic resources may be reappearing in the emerging debate over the management of human genetic resources.

Many of the issues that have challenged the plant genetic resources community over the past two decades, including the need for intergovernmental involvement with respect to the collection, storage, exchange, benefit-sharing and IP aspects of plant germplasm, also arise with regard to human genetic diversity — albeit with more profound moral and ethical considerations.

Controversy over the collection and patenting of human genetic material is not new. In 1993 the Human Genome Diversity Project, an informal consortium of universities and scientists in North America and Europe, proposed to collect human DNA samples from hundreds of so-called 'endangered' indigenous communities around the world. Many indigenous peoples' organizations protested vigorously, asking: Will profits be made from the genes of poor people whose physical survival is in question? Who will have access to stored DNA samples, and where will these collections be located? What benefits, if any, will accrue to the indigenous peoples from whom DNA samples will be taken?

In March 1995 the US patent office issued a patent on a cell line containing unmodified DNA from a Hagahai tribesman in Papua New Guinea. Indigenous peoples' organizations vocally denounced the patent as a threat to human dignity and a violation of human rights. The controversy, generated by scores of indigenous peoples' organizations, together with civil society organizations and governments, eventually caused the US government to 'disclaim' the Hagahai patent in October 1996.

Commercial trade in human tissue is accelerating. Scientists, in both the public and private sector, are collecting human DNA samples from rural and urban communities across the globe. Of particular interest to genetic researchers are populations that are genetically homogeneous, or those that exhibit a genetic predisposition to an inherited disease. After pinpointing the location of so-called 'disease genes' genomic companies and their pharmaceutical partners hope to develop commercial products such as diagnostic tests and therapies that are based on proprietary human genes. That quest has taken gene prospectors to remote locations such as Tristan da Cunha in search of asthma genes, to Kosrae in Micronesia in search of obesity genes, and to Tibet in pursuit of high-altitude genes, just to name a few.[82]

In early 1998 the prospect of nationwide collection and commercialization of human DNA made headlines when Hoffman-La Roche (Switzerland) and DeCode Genetics Inc. (Iceland) signed a $200 million collaborative research contract to identify disease genes based on studies of Iceland's relatively isolated and strikingly homogeneous population.[83] DeCode's goal is to amass

See p. 111 *et seq.* for the Notes

the world's most comprehensive collection of genealogical family data for studying the genetic causes of common diseases. The company says that its studies could lead to new diagnostic tests and drugs for inherited diseases — which would be made available free to Icelanders if the research leads to a new therapy.[84] The Icelandic situation has become an international test case for many of the ethical and intellectual property issues surrounding collection and commercialization of human DNA.[85] Despite opposition by growing numbers of Iceland's scientific and medical community, a bill was passed by the Icelandic parliament on 17 December 1998 that gives DeCode Genetics the right to collect current and retrospective medical information from Iceland's 270 000 inhabitants into a centralized, comprehensive database.[86] The new law gives DeCode Genetics exclusive rights to the commercial exploitation of genetic information for 12 years.

A vocal minority of Iceland's scientific and medical community, including the Icelandic Medical Association, the Association of Icelanders for Ethical Science and the Icelandic Mental Health Alliance oppose implementation of the law and are advising doctors and their patients to refuse participation in the collection of DNA samples.[87] Opponents believe that the bill violates principles of privacy and informed consent and they object to a single company gaining exclusive rights to a valuable scientific resource. For example, the law allows only for individuals to opt out of the database, but does not require any other form of consent. Although the database is supposed to be confidential and anonymous, critics charge that personal information can be deciphered and that computer security measures proposed by the company are not adequate to insure confidentiality.

Crucible Recommendation 5

Protecting human genetic diversity
The Crucible Group notes that human genetic material is being collected worldwide in the absence of intergovernmental oversight, and without consistent regulations concerning the collection, exchange and use of human genetic diversity and the protection of human subjects.

The Group recommends that all aspects of the conservation and utilization of human genetic diversity be governed, monitored and reviewed by governmental or intergovernmental agencies, as appropriate, with the full, informed consent and participation of all human subjects involved, especially indigenous peoples.

Bioethics and societal choices: who will decide?

By the close of the 20th century, humankind had acquired the power to transform the processes of all living species, including its own. The potential benefits of these powers can be exciting and the perils ominous. Consider, for example, the recent announcement that scientists have successfully produced cultures of embryonic stem cells.[88] This breakthrough offers the

See p. 111 *et seq.* for the Notes

potential to grow any type of human tissue and may eventually be used to repair damaged hearts, blood vessels or brains. The very same week, the UK's *Sunday Times* reported that scientists can theoretically engineer deadly biological organisms to produce 'ethno-bombs' that are capable of targeting human victims by ethnic origin.[89]

Given the dizzying pace of technological advancements in genetics and biology, it is not surprising that society is grappling ever more urgently with the social, ethical and legal implications of humankind's ability to decipher and control the genetic blueprint of life. Opinions differ sharply on the implications of new biotechnologies, but nearly everyone agrees that advances in technology are taking place at a rate far faster than social policies can be devised to guide them, or legal systems can evolve to address them.

There is growing recognition worldwide that the development of scientific knowledge must be accompanied by public debate on societal choices and the informed participation of citizens. 'Bioethics' attempts to identify the social and cultural implications of breakthroughs in life sciences, to anticipate its applications, and to ensure that progress in the life sciences benefits humanity as a whole.[90] Bioethics acknowledges that there is a distinction between what is scientifically possible and ethically acceptable. What is good for society, what is equitable, and what is safe? Who will decide? These are among the questions that are fueling debate on the applications of biotechnology to health, agriculture and human development.

At the intergovernmental level, the United Nations Educational, Scientific and Cultural Organization (UNESCO) created the International Bioethics Committee in 1993 as the world's only international body to study the implications of human genome research and genetic engineering. In November 1997, UNESCO adopted a non-binding Universal Declaration on the Human Genome and Human Rights, the first international text on the ethics of genetics research. Since 1993, a growing number of countries have established national bioethics advisory committees to examine bioethics and to provide guidance to national governments. In 1998, the CGIAR adopted a set of ethical principles to guide its work on genetic resources.

The Crucible Group recognizes and appreciates the vital contribution of ethical debates. There is concern, however, that the appointment of expert panels and commissions devoted to bioethics should not become a substitute for broad public debate and participation in the review and assessment of new technologies.

See p. 111 *et seq.* for the Notes

Changes in Molecular Bioscience: What Impact on Society and Biodiversity?

Introduction

What has changed in molecular sciences in the past five years, and how does that influence the way society thinks about, uses and values biodiversity? Biological knowledge continues to expand rapidly. By one estimate, the ability to identify and use genetic information is doubling every 12 to 24 months.[91] Scientific and technical breakthroughs in the biological sciences not only change the practice or interpretation of science, they often have profound implications for society.

The industrialization of a gene-based strategy to predict, understand and manipulate biological organisms for commercial agriculture and human health is being hailed as the engine that will drive economic development in the 21st century. In the words of one industry spokesman: 'Automation has allowed us to put biological discovery on an assembly line.'[92] The potential benefits of sequencing genes and identifying their functions are easy to predict. More difficult to answer are questions such as: Who will have access to the technology, and who will determine priorities for using the information? Some members of the Crucible Group are concerned that high-profile technology projects risk overshadowing more basic priorities for human health and agriculture.

This chapter briefly introduces some of the recent breakthroughs in molecular science and technology relating to plants and livestock, as well as humans, in the field of agriculture and human health. This is by no means an exhaustive survey, but an attempt to identify landmark scientific and technological breakthroughs (primarily in molecular science). It is important to note that, in many cases, the same technologies (i.e. genomics) are increasingly applied to humans, plants and animals alike.

Members of the Crucible Group wish to emphasize that breakthroughs in science and technology are not confined to high-tech laboratories or white-coated scientists. Technologies to develop, utilize and conserve genetic resources are derived not only from formal sector institutions, but also from local ecosystems, local knowledge and traditional practices. While this chapter focuses on highly-visible landmark achievements in molecular sciences and technology, the Crucible Group notes that one of the most important changes in the past five years has been increased recognition, in some circles, for the role of farmers and local communities in plant breeding/selection, conservation and use of biological diversity (see discussion in previous chapter).

See p. 111 *et seq.* for the Notes

Mammalian cloning: Dolly debuts

A sheep named 'Dolly' tops the list of recent scientific breakthroughs. Mammalian cloning became a riveting reality in February 1997 when the Scotland-based Roslin Institute unveiled Dolly — a lamb cloned from a single cell of an adult sheep. Within days of the announcement, 'cloning' became a global household word, and scientists and historians began to re-write outdated text books. Dolly's public profile is so high that the Roslin Institute recently applied for trademark protection of her name and image.[93]

Before the Roslin Institute announced Dolly's birth, patent applications were filed on the technique used to clone her. PPL Therapeutics, one of the Roslin Institute's for-profit pharmaceutical partners, saw its stock value jump overnight. But the Dolly-related patent claims proved controversial because the applications filed by the Roslin Institute at the World Intellectual Property Organization (WIPO) were not limited to a technique for cloning farm animals — they included all mammals — and did not exclude humans. If national patent offices around the world grant patents on the technique to clone Dolly, would they be implicitly accepting techniques to clone human beings, or even the morality of those techniques themselves? If so, are major social issues being determined in national patent offices in the absence of informed public debate? (It is important to note that a patent has nothing to do with the ability to commercialize or market a new technology; it is a right to prevent others from using or selling the technology without authorization.)

In July 1998 scientists under the direction of Ryuzo Yanagimachi at the University of Hawaii provided the first scientific report confirming that the technical feat of cloning from adult mammalian cells could be replicated, putting to rest growing speculation that Dolly was nothing more than a laboratory fluke or a fake.[94] The University of Hawaii scientists produced three generations of cloned mice, more than 50 in all. The Hawaiian team has also filed for patents on the novel aspects of its cloning technique.[95] The same month, Japanese researchers announced the birth of two cloned calves. In April 1999 Genzyme Transgenics announced the arrival of three cloned, transgenic goats, one of which is engineered to produce a human protein in its milk.

While ethicists and policy-makers continue to struggle with the unsettling implications of mammalian cloning, the technique has advanced — over a period of two years — from stunning front page news to what *Time* magazine has labeled 'almost a mundane laboratory practice'.[96] However, mammalian cloning is still relatively inefficient and will require considerable refinement before it becomes a commercially viable technique.[97]

The FAO has concluded that somatic cloning technology offers a potentially valuable tool to save domestic animal breeds in danger of extinction. In late 1997 FAO elaborated a protocol to collect and store samples of animal tissue in the expectation that cryopreservation of somatic tissues will eventually enable scientists to recreate endangered breeds.[98] In New Zealand, scientists recently revealed the successful cloning of a calf from the last surviving female member of an ancient herd of short-horn cattle.[99]

In the final days of 1998 the spectre of human cloning made headline news when a South Korean physician reportedly conducted the first-ever human-cloning experiment.[100] At an infertility clinic in Seoul, Lee Bo Yon produced a four-cell embryo from genetic material extracted from a 30-year-old woman. The experiment was terminated because a Korean code of conduct forbids the insertion of a cloned human embryo into a womb. In reaction to news of the Korean experiment on human cloning, Dr Mary Lake Polan of Stanford University told the *Wall Street Journal*, 'If there is a market for it, and it is technically possible, then someone will do it.'[101]

In the US, federal laws prohibit government funding of research on human embryos, but independent scientists are not restricted from human cloning experiments. In December 1998 a scientific advisory committee in the UK proposed that the government allow human cloning for medical research but not to produce babies.[102] In theory, so-called 'therapeutic cloning' could lead to the production of spare body parts for transplant surgery that would be genetically identical to the patient's. However, in mid-1999 researchers revealed that DNA in Dolly's cells are typical of a much older animal, a finding that could have implications for the commercial-scale production of cloned animals and their use in transplantation medicine.

DNA sequencing accelerates

Genomics — the science of identifying the entire set of genes of living organisms — is revolutionizing biological sciences and has become a driving force in mergers and divestments involving many of the world's largest corporations.[103] Genomics technologies are now being used by public and private researchers to decipher the genetic blueprint of humans, plants, animals and micro-organisms. As the efficiency of DNA sequencing technology accelerates, genomics milestones are being reached far ahead of schedule.

In 1995, a commercial genomics company announced that it had sequenced the entire genome of a living organism, the bacterium *Haemophilius influenzae*, and that it had filed for broad patent claims on the medical uses of the organism's bacterial proteins.[104] 'It is the first time the entire genetic content of a free living organism has been deciphered', said William Haseltine, chief executive of Human Genome Sciences (HGS).[105] By mid-1997, HGS had sequenced the entire genome of three additional bacterial pathogens. Although patents have not yet been issued, HGS's claims include the development of diagnostics, vaccines and antibiotics related to their proprietary genomics information.[106] Today, whole-genome sequencing of microorganims is commonplace. By the end of 1997, more than 50 microbial genome projects were underway worldwide.[107] As of mid-1998, 14 bacterial

See p. 111 *et seq.* for the Notes

and archaeal genomes had been completely sequenced.[108] Scientists announced completion of the sequencing of the first animal genome, *Caenorhabditis elegans* (a nematode worm), on 10 December 1998. *C. elegans*, only one millimetre in length, is the first multi-cellular organism whose entire genetic sequence — some 100 million base pairs — is known.[109]

Human Genome Project

The Human Genome Project, a 15-year, $3 billion project supported primarily by the US government and British partners, was launched in 1990 to map the entire human genome, the 80 000–100 000 genes that exist within our DNA. Advocates of the Human Genome Project describe it as 'more important than putting a man on the moon or splitting the atom'.[110] Sequencing the entire human genome requires the identification of more than 3 billion nucleotides or base pairs (the molecular letters that make up our genetic code). It also involves arranging the molecular letters in a precise order and learning how to read them.

The process of sequencing DNA is now faster and cheaper than anyone imagined possible five years ago. For example:

- In the mid-1970s it would take a laboratory two months to sequence 150 nucleotides.[111] Today, one commercial genome firm has the tools to sequence 11 million letters a day.
- It took 1 000 scientists ten years to decode a yeast genome using the first generation of high-tech gene sequencers. Using today's state-of-the-art computers scientists could complete the same job in one day.[112]
- The cost of DNA sequencing has dropped from about $100 per base pair in 1980 to less than a dollar today, and experts predict it will be less than a cent by 2002.[113]

The Human Genome Project was conceived as an international, public sector initiative, a project too massive in scope and too expensive for any single country or company to undertake. With the advent of faster, cheaper sequencing technologies, the race to map the human genome now faces stiff competition from the private sector.[114] In May 1998 a new commercial venture announced that it would start and essentially complete the sequencing of the human genome in 2001, four years ahead of the US government's target date of 2005. The company, a joint venture between Perkin-Elmer and the US-based Institute for Genomics Research (TIGR) claims that the sequencing capacity of the company's state-of-the-art equipment far exceeds the total sequencing capacity of all existing genomics laboratories in the world.[115] The new company's goal is to become the 'definitive source of genomics and associated medical information'.[116]

Spurred by competition from the private sector, the Wellcome Trust of London, the world's largest medical philanthropic organization, announced in May 1998 that it would double the money it contributes to the UK-based Sanger Centre, enabling its biologists to sequence one-third of the human

See p. 111 *et seq.* for the Notes

genome in partnership with the Human Genome Project. Bolstered by international support, the Human Genome Project announced in September 1998 that it would move up by two years, to 2003, its target date for completing the sequencing of the human genome.[117] Appropriately, 2003 marks the 50th anniversary of the discovery of the double helix.

Human gene patenting

William Haseltine of Human Genome Sciences claims that by 1995 his company had already isolated 'greater than 95 percent of all human genes'.[118] (Only about 3% of the DNA in the human genome codes for genes, and the rest is considered 'filler'.) Many commercial ventures are concentrating on the 'gene-rich' regions of the human genome, ignoring the non-coding DNA. On 17 September 1998 HGS announced that patent applications published under the auspices of the Patent Cooperation Treaty include claims on a total of 476 full-length human genes.[119] According to HGS, each of the genes described in the patent applications represents a newly described human gene in the form of a corresponding complementary DNA (cDNA), the complete protein coding text of each gene and potential medical uses.

The furious pace of discovery in the field of genomics is reflected in the growing number of patent claims related to partial gene sequences or ESTs (expressed sequence tags). In 1991, the US Patent and Trademark Office had applications pending on 4000 EST sequences. In 1996, there was a total of approximately 350 000 EST sequences to be examined, and as of September 1998, there were applications pending on over 500 000 EST sequences.[120]

The patenting of partial gene sequences, or ESTs, is controversial. Many members of the scientific and patent community, including the US government's National Institutes of Health object to the patenting of ESTs. How, they ask, can standard patent criteria (novelty, non-obviousness and utility) be met in a case where the function of a partial gene sequence (the protein it encodes) is not even known? Many view it as a distortion of the patent system to allow patents on information that can be decoded by computers and does not appear to involve an inventive step. There is concern that claims on partial gene sequences may preclude future patenting of a full-length gene that contains an already patented sequence.

In November 1998 California-based Incyte announced that it had received the first patent on 44 ESTs.[121] The company's self-described aim: 'Our goal is now to have sequenced, mapped and filed for intellectual property on the novel and most commercially relevant genes by the second half of the year 2000.'[122]

In September 1999 it was reported that the UK and US governments were negotiating an Anglo-American agreement that seeks to release all publicly-funded research on human genes without claiming patents.[123] The goal of the proposed agreement is to ensure that the benefits of human genomics discoveries are made freely available worldwide.

See p. 111 *et seq.* for the Notes

Artificial human chromosomes

In 1997 researchers at Case Western Reserve Medical School in Ohio (US) announced the creation of a promising new gene carrier: a human artificial chromosome (HAC) that behaves just like a natural one in cultured human cells. [124] The human artificial chromosome is replicated and passed along with every cell division.

HACs provide potentially powerful tools for the analysis of chromosomal functions and also for the cloning of large DNA fragments. They could someday be used to introduce large fragments of DNA into cells or whole animals in a stable form.[125] Scientists are attempting to create HACs that contain specific human genes so that they can study how they function in cell cultures.

Scientists are also experimenting with the stability and expression of artificial chromosomes in hamster, mouse and chicken cells. While the use of HACs in human gene therapy remains distant, there is speculation that, once perfected, artificial chromosomes could become a vector for delivering complex, custom-made genetic programs into human embryo cells.[126] The once-unthinkable notion of human genetic engineering thus becomes technically feasible.

On the horizon: pharmacogenomics

Advances in the field of genomics are adding new words to the scientific vocabulary. 'Functional genomics' involves analyzing the role of genes in disease. The next step, dubbed 'pharmacogenomics', will use advanced genetic tools to compare how genetic information varies from individual to individual.[127] Many public and private sector researchers are now cataloguing minute genetic variations — single nucleotide polymorphisms, or SNPs — that may correlate with an individual's susceptibility to disease or response to drugs. Technology for the rapid screening of SNPs will someday make it possible to obtain a unique and precise genetic profile for each individual. Standard gene sequencing technology would require at least two weeks and $20 000 to screen a single patient for genetic variations in 100 000 SNPs. But one commercial DNA chipmaker says that it is developing a technology that can screen 100 000 SNPs in a patient's genome in several hours, for just a few hundred dollars.[128]

Precise genetic profiling could allow drug companies to customize prescription medicines, and to know, in advance, if an individual's genetic make-up would cause an adverse reaction to a particular drug. The prospect of genetic profiling also raises far-reaching ethical and legal issues relating to the potential misuse of genetic information, (i.e. violations of privacy and informed consent, genetic discrimination from insurance companies, employers, etc.).

See p. 111 *et seq.* for the Notes

Advances in drug research and discovery

Jim Niedel, executive director of research at Glaxo Wellcome, describes the genomics era as the beginning of the third generation of drug research. According to Niedel, the first generation, which started about 100 years ago, was based on chemistry and serendipity. The second, beginning in the 1950s, was based on biology and empiricism. The third generation depends on skilled professionals using genetics, robotics and informatics.[129]

The newest, state-of-the-art equipment for 'high-throughput screening' (that is, automated testing of a large number of chemicals against disease targets) is capable of preparing samples for up to 100 000 screening tests per day, the equivalent of a month's worth of manually-prepared samples.[130] High-throughput screening uses robots to test the actions of thousands of compounds against a molecular disease target.

After screening provides a promising drug candidate, combinatorial chemistry is the next step. With robotic assistance, chemists can compose thousands of variations on the original chemical — producing a family of molecules related to the original candidate. The process has the potential to vastly accelerate drug discovery. For example, a laboratory at Glaxo Wellcome took just one month in 1997 to sift 150 000 chemical processes for the best way to build a class of drugs for respiratory, neurological and viral disorders. The new era of drug discovery requires the technical capacity to digest massive quantities of data. The pharmaceutical company SmithKline Beecham had only two bioinformaticians four years ago, yet nowadays the company employs 70 of them.

New methods for identifying, quantifying and controlling the active components of plants also provides new opportunities for the development of herbal medicines. Over 80% of the world's population rely on local health practitioners and traditional medicines for their primary medical needs.[131] The worldwide market for herbal medicines is an estimated $12.6 billion. A biotechnology company in the United States is developing a new technology that aims to develop clinically tested and regulated prescription and non-prescription drugs derived from herbal medicines. The company, PharmaPrint, creates a 'fingerprint' of the herb that can be used to identify the quantity and bioactivity of each active component in an herbal medicine. Many people assume that traditional herbal remedies are non-patentable because the knowledge exists in the public domain.[132] But the PharmaPrint process reportedly replicates the chemical compounds in the herb, 'allowing for a chemical product that can be patented and that can be standardized for clinical development and commercialization'.[133] Herbal medicines now in clinical development in the US include, for example, mistletoe, black cohosh, St. John's wort, saw palmetto, valerian, milk thistle, agnus castus, and ginkgo biloba.[134]

See p. 111 *et seq.* for the Notes

Crop genomics research accelerates

Scientists are using advanced genomics as a means of identifying, mapping and understanding the expression of crop genes, and their link to agronomically important traits. The goal is not only to construct genetic maps of plant species, but also to link the genetic structure of the plant with its protein activity.[135]

Since 1996, virtually every major seed company has invested in plant genomics research. Driven by the increased efficiency of genomics technology and fierce competition among major agrobiotechnology firms, investments in crop genomics accelerated dramatically in 1998 (see Table 2).[136] Private sector investment in crop genomics dwarfs public spending in this arena.

In 1996, Pioneer Hi-Bred launched a $16 million maize genomics program with Human Genome Sciences. DuPont (which announced in early 1999 that it would acquire Pioneer Hi-Bred) has a major R&D focus on plant genomics. According to Anthony Cavalieri of Pioneer, the company has identified pieces of DNA matching over 350 000 maize genes, accounting for over 80% of the maize genome.[137]

In September, 1997 Monsanto and Millennium Pharmaceutical announced a five-year partnership worth up to $218 million to identify patentable crop

Table 2: Recent agricultural genomics deals

Company/Institute	Partner	Date	Action
AgrEvo (Germany)	Gene Logic	1998	3-yr., $45 million genomics research alliance
Dow (USA)	Biosource Technologies	1998	3-yr. genomics research alliance
DuPont & Pioneer (USA)	CuraGen	1998	$5 million per annum expansion of plant genomics research alliance
DuPont (USA)	Lynx Therapeutics	1998	5-yr., up to $60 million. Focus on maize, soybeans, wheat & rice
Genoplante— French Genome Initiative (France)	Public/private alliance involving Rhone-Poulenc, Biogemma, Sigma/Serasem, Florimond Desprez, INRA, CIRAD, ORSTOM and French universities	1998	Focus on genomics in European crops
Monsanto (USA)	Incyte Pharmaceuticals	1998	Broad access to Incyte's gene expression technology
Monsanto (USA)	GeneTrace	1998	$17.2 million, plant & animal agricultural genomic technology
Novartis	Novartis Ag. Discovery Institute	1998	10-yr., $600 million plant genomics institute
NSF Plant Genome Research Project (US govt.)	University of Missouri (USA)	1998	$11 million for maize genomics research
Zeneca (UK)	John Innes Centre and Sainsbury Laboratory (UK)	1998	10-yr., $80 million for advanced genomics and wheat
Zeneca Agro (UK)	Alanex	1998	3-yr. agreement for screening Alanex's compound library

See p. 111 et seq. for the Notes

genes using genomics technologies. The exclusive deal is not limited to a single crop or geographic location, instead it covers all crop plants in all countries.

In July 1998 Novartis announced that it would spend $600 million over ten years to establish the 'Novartis Agricultural Discovery Institute,' a new in-house effort dedicated to plant genomics research. The company says it will be the world's biggest crop gene mapping project. The California-based institute will employ about 180 scientists.

Genetic use restriction technologies

On 3 March 1998 Delta & Pine Land Co. in Mississippi and the US Department of Agriculture (USDA) announced that they had received US patent 5 723 765 on a new genetic technology designed to prevent unauthorized seed saving by farmers.[138] The patent is a prototype of similar techniques that are being developed with the aim of genetically altering second generation seed to prevent germination. The developers of the technology refer to it as a 'technology protection system' or 'genetic use restriction technology' (GURT). It is popularly known as the 'Terminator'. Developers of terminator-type technologies indicate that it will be at least four years before seeds incorporating the 'suicide trait' are available for commercial sale.

GURT or Terminator is not a single technique being developed by a single company. Over 30 patents have been issued to 13 institutes (public and private) that describe techniques for controlling seed germination, and/ or the use of 'inducible promoters' to activate traits or performance of genetically engineered plants. Inducible promoter systems enable plant genes or traits to be genetically triggered by the application of an external chemical catalyst. In the future, farmers would theoretically be able to activate or deactivate genetic traits such as germination or insect resistance by applying a prescribed chemical to their seed. Critics warn that the development of these technologies will dramatically increase farmers' dependence on agrochemical companies and their proprietary inputs.[139] Proponents believe that farmers will be afforded more options, if given the freedom to choose whether or not and under what circumstances to trigger value-added traits in genetically engineered plants.[140]

Virtually all major seed and agrochemical corporations are conducting research and development of GURTs. If commercially viable, genetic seed sterilization could, some believe, have far-reaching and negative implications for farmers and food security.

According to a USDA spokesman, Willard Phelps, the goal of the USDA's new technology is 'to increase the value of proprietary seed owned by US seed companies and to open up new markets in Second and Third World countries'.[141]

The president of Delta & Pine Land, Murray Robinson, told a US seed

See p. 111 *et seq.* for the Notes

trade journal that his company's seed sterilizing technology (co-owned with the USDA) could be used on over 405 million hectares worldwide (an area the size of South Asia), and that it could generate revenues for his company in excess of $1 billion per annum.[142] Robinson says that his company's newly patented technique will provide seed companies with a 'safe avenue' for introducing their new proprietary technologies into giant, untapped markets such as China, India and Pakistan.[143] The USDA and Delta & Pine Land Co. have indicated that they will apply for patent protection in 87 countries worldwide, including many nations throughout Africa, Asia and Latin America.

A substantial number of civil society organizations warn that GURTs threaten food security and agricultural biodiversity, especially for the poor, because if widely adopted this technology could restrict farmer expertise in selecting seed and breeding locally adapted varieties.[144] Over 1.4 billion people — primarily resource-poor farmers in the South — depend on farm-saved seed and seed exchanged with farm neighbors as their primary seed source.[145] Many CSOs have called for a global ban on the technology, which they view as an immoral technique that will rob farming communities of their age-old right to save seed.

A scientific panel convened by the Conference of Parties to the CBD recommends that intergovernmental bodies such as the FAO, in close collaboration with UNESCO and the United Nations Environment Program (UNEP), further study the implications of GURTs on the conservation and sustainable use of agricultural genetic resources, and identify relevant policy questions that need to be addressed.[146]

Proponents of the gene protection technology claim that, if the private sector is able to protect its research investment, it will spur investment in plant breeding for many of the world's most important crops. Proponents believe that the gene protection method, if it can be made to work effectively, offers a new tool for controlling involuntary outcrossing, which will ultimately protect crop integrity and preserve global biodiversity.[147] Genetic seed sterilization could also be used to prevent early sprouting in the field, a major problem in tropical countries when the harvest season is wet. Advocates also point out that farmers will always be free to choose whether or not to buy gene-protected seeds — and will not do so unless such seeds offer them a clear advantage over fertile seed that will compensate for their higher cost. In the words of one industry advocate, traditional farming practices such as seed-saving can put resource-poor farmers at a distinct disadvantage: 'The centuries old practice of farmer saved seed is really a gross disadvantage to third world farmers who inadvertently become locked into obsolete varieties because of their taking the 'easy road' and not planting newer, more productive varieties.'[148]

Genetic seed sterilization is a subject of controversy and debate worldwide.

- In May, 1998 the Conference of the Parties to the Convention on Biological Diversity (COP IV) recommended that the precautionary principle be applied to the use of new technology for the control of plant gene expression. COP IV directed its Subsidiary Body on Scientific, Technical and Technological Advice to consider the technology's impact on the conservation and sustainable use of biodiversity.
- India's agriculture minister Som Pal told the Indian parliament in August 1998 that he had banned the import of seeds containing the terminator gene because of the potential harm to Indian agriculture.[149]
- At its annual meeting in October 1998 the CGIAR adopted a policy stating that it would not incorporate into its breeding materials any genetic systems designed to prevent seed germination.[150]
- In February 1999 a spokesman for Zeneca declared: '[The company] is not developing any system that would stop farmers growing second-generation seed, nor do we have any intention of doing so.'[151]
- In June 1999 the president of the Rockefeller Foundation advised the biotechnology industry to 'disavow' the use of the terminator technology to produce seed sterility.[152]
- In April 1999 the Monsanto Company announced that 'concerns about gene protection technologies should be heard and carefully considered before any decisions are made to commercialize them'.[153] In October

Viewpoint box: Terminating Terminator?

Members of the Crucible Group do not agree on whether or not terminator technology could or even should be banned. There are different viewpoints on whether or not the public morality clause of TRIPs (Article 27.2) could be used to exclude terminator technology.

GURT technology is not in conflict with the public morality clause of TRIPs.
It would be unfair to exclude a technology as a whole from patent protection, just because it may be abused for immoral purposes or may have negative side effects. Apart from that, the GURT technology can obviously be used for purposes that are perfectly moral — even beneficial — and do not violate the *ordre public*. Furthermore, to deny farmers the opportunity to make their own choice is restrictive paternalism, even if with good intentions.

The terminator technology is immoral
The technology is intrinsically immoral and has been developed for no other purpose than to prevent farmers from replanting seeds. Given that the technology may even affect farmers who never used the terminator seeds, the technology should be banned. It is a positive sign that two major agrochemical corporations have made a commitment not to commercialize sterile seed technologies. However, we can't depend on the goodwill and charity of corporations that may be acquired by another company next month. In order to discourage the development of similar technologies, governments should also ensure that their patent laws do not set any incentive to develop similar technologies. The monopoly control afforded by terminator technology goes far beyond patents and threatens national sovereignty. A patent is a time-limited, legal monopoly granted by a government in exchange for societal benefits. In the case of the Terminator, the biological monopoly is not time-limited, and is not necessarily approved by national governments.

See p. 111 *et seq.* for the Notes

1999 Monsanto's chief executive Robert B. Shapiro made a public commitment not to commercialize sterile seed technologies.[154] The company did not rule out the future development and use of genetic trait control as a means of gene protection.

Crucible Recommendation 6

Genetic use restriction technology
The Crucible Group acknowledges the scientific achievement in developing GURTs. While the social and economic implications of this technology for the food security of rural communities in developing countries have yet to be studied, some public and private institutions have made commitments not to commercialize or use sterile seed technology. Irrespective of the social and economic impacts, the Crucible Group recommends that the technology not be used in released varieties where its primary purpose is to prevent seed-saving among resource-poor farmers in developing countries.

Clonal plant reproduction by apomixis

Apomixis is a natural, asexual type of reproduction in which plant embryos grow from egg cells without being fertilized by pollen. Apomixis offers a means of cloning plants through seed. The progeny are genetically identical to the mother plant. Apomictic seed is genetically uniform from generation to generation (unlike normal sexual hybrids or open-pollinated varieties).

In contrast to the gene technology protection systems (terminator technology) described above, which are designed to prevent farmers from saving second generation seed, apomixis technology has the potential to dramatically expand and de-centralize plant breeding opportunities, especially for resource-poor farmers. In theory, apomictic hybrid seed could offer tremendous benefits to resource-poor farmers because desirable traits could be maintained indefinitely, with no loss of hybrid vigour, and farmers would be able to save their hybrid seed for replanting year after year. Apomixis technology could offer fast, flexible and low-cost plant breeding strategies that would be responsive to locally-targeted crop breeding needs.

Apomixis occurs naturally in many plant species and wild relatives of some crops. The challenge is to introduce the trait for apomixis into sexually propagated crops such as rice, wheat, millet, sorghum, etc. Plant breeders and molecular biologists have successfully transferred the genes that confer apomixis from a wild grass species, *Tripsacum dactyloides*, to maize. Maize is the first sexual species successfully transformed into an apomictic form.[155]

Who is working on apomixis? Many public and private agricultural researchers in both developing and industrialized countries are conducting research on apomixis, and over two dozen patents have been issued related to apomixis technology.[156]

Virtually all major life sciences corporations (multinational seed and agrochemical corporations) have an interest in apomixis research, particularly because of its potential to reduce the cost of hybrid breeding programs.[157] Once a superior variety is created by the combination of inbred lines, the apomictic hybrid plant and its genetically uniform offspring could produce seeds asexually more conveniently than inbred lines. Using apomixis to produce hybrid seeds, companies could drastically cut costs associated with maintaining inbred lines, including land and labour-intensive practices such as detasseling to prevent cross-pollination.

For seed companies, hybrid technology is a form of built-in proprietary protection. However, apomixis technology could undermine the proprietary protection afforded by traditional hybrids because farmers would be able to save and sow apomictic hybrids. As a result, seed companies are interested in combining new developments in genetic seed sterilization with apomixis. If commercial seed firms can successfully combine the benefits of apomixis (the ability to mass-produce low-cost clones) with genetic seed sterilization this will eliminate the ability of farmers to save and re-use seed from apomictic varieties. However, it remains to be seen if this is technically feasible.

There is concern that mass production of low-cost clonal varieties could promote genetic uniformity and crop monoculture. The introduction of genetically uniform cultivars could unintentionally reduce genetic diversity in agriculture, if widely introduced in areas where farmers grow traditional crop varieties. Proponents of apomixis respond to these concerns by pointing out that apomixis will also permit the rapid development of new, resistant varieties on a more regular basis. The simplicity and low cost of apomictic breeding would encourage the introduction of a wider range of varieties that could be uniquely suited to a particular micro-environment, and thus encourage genetic diversity in farm communities.

Who will benefit from apomixis? Apomixis technology has the potential to profoundly influence farming systems and revolutionize plant breeding worldwide. Ultimately, both the private and public sectors will play important roles in developing apomixis technology for potential use in a wide range of farming systems. The Crucible Group notes that the benefits of apomixis technology for resource poor farmers depends largely on the ability of farmers themselves to gain access to, manage and control the use of apomixis, and to experiment/innovate with the crossbreeding of locally adapted varieties, in partnership with National Agricultural Research System (NARS) and the International Agricultural Research Centres (IARCs) of the CG system.[158]

See p. 111 *et seq.* for the Notes

Part Two: Outstanding issues
Access and exchange, knowledge, and innovation

Policy primer

Major changes in the policy environment

In the last five years the international policy environment has weathered its own versions of global warming and acid rain as new intergovernmental conventions and agreements, only theoretical five years ago, have come into force. Not surprisingly in this interrelated world, diverse global or regional agreements on the management of intellectual property, biological diversity, trade and human rights may all have implications for the conservation and use of genetic resources — including their control, ownership, and access/exchange arrangements. While there is universal acceptance that the landscape has changed substantially, there is no agreement as to which of the changes can be considered positive or destructive. Among the major new developments:

- 1993 — the Convention on Biological Diversity (CBD) came into force.
- 1994 — the CGIAR's International Centres signed agreements placing most of their germplasm collections 'in trust' under FAO auspices.
- 1995 — the World Trade Organization came into being with its chapter on Trade-Related Aspects of Intellectual Property Rights (TRIPs) requiring protection for plant varieties.
- 1996 — the Leipzig Global Plan of Action for Plant Genetic Resources for Food and Agriculture (PGRFA) was adopted (though it has not been *fully* implemented).
- 1998 — the 1991 Act of the Union for the Protection of New Varieties of Plants (UPOV) Convention entered into force, closing the door for new parties to join the 1978 accord.
- 1999 — the FAO Commission on Plant Genetic Resources for Food and Agriculture continues to re-negotiate its International Undertaking and, as part of this, Farmers' Rights.

The broad changes in science, technology and the environment described in Part One play out in a variety of interrelated intergovernmental fora. Policy-makers tend to frame and understand policy issues as they are negotiated and monitored through each UN or Bretton Woods institution. In fact, however, these issues are not so readily compartmentalized. There is a real danger of losing track of the great themes and trends amidst the minutiae of international conventions. In Part Two, the Crucible Group considers the central policy issues in three wide topic fields: germplasm access and exchange; knowledge conservation and formation; and, finally, innovation management. In each of these fields, the Group considers the most significant intergovernmental negotiations. Since developments in one forum influence decisions in another, readers will find themselves reminded to cross-reference many specific points and to be aware that, for example, a position taken by trade negotiators in one convention could enhance or damage national environmental policies being debated in another forum.

See p. 111 *et seq.* for the Notes

By dividing their discussions into the three fields (Access/Exchange, Knowledge, Innovation), the Group felt that it could further assist national policy-makers in formulating legislative choices. As already noted, the second volume of this Crucible Report gives detailed consideration of a number of legal and policy options that are either operational or under development in different parts of the world, or are theoretical possibilities that are not under active consideration. Readers may wish to refer to similarly named parts of Volume 2 (and vice versa) as they go through this report.

Obviously, it is increasingly difficult for national policy-makers to identify issues and options. A study currently being undertaken by the Agricultural Research Group of the World Bank — specifically, the Development Agricultural and Extension Group (ESDAR) — is considering how governments are facing the predicament. In studies conducted with the Swedish International Development Cooperation Agency (Sida), the two organizations appear ready to conclude that governments require much greater internal coordination in order to have an organized and rational strategy that applies logically in every international forum.[159] The confusion between fora is further exacerbated by the often uneven playing field that typifies many intergovernmental negotiations. For example, at the beginning of the 1990s, as the world community reached critical decisions on trade, the environment, intellectual property, new technologies and genetic resources, more than 90% of the databases on Africa were to be found solely in industrialized countries — not in Africa.[160] Not only was the information

---

Crucible Recommendation 7

Ensuring effective participation in — and policy harmonizing among — intergovernmental fora

The Crucible Group notes that the development of consistent and complimentary policies in various intergovernmental fora has sometimes been frustrated by financial barriers to effective participation for some countries and regions and by inconsistent policy preparation at the national level in other countries and regions. In order to avoid serious errors and contradictions, the Group recommends that:

- all parties strive to guarantee a genuinely 'level playing field' for all governments with respect to information access, preparation periods, and participation numbers in all associated negotiations;
- while acknowledging the primacy of government to identify and represent national goals and policies, governments establish formal consultative processes through which all of the public, private, and civil society organizations associated with the range of issues involved are able to advise and monitor their governments' activities in each forum. In particular, with respect to genetic resources for food and agriculture, the full participation of all stakeholders including farmers and plant breeders as well as indigenous people and local communities is essential to good governance.

See p. 111 *et seq.* for the Notes

not easily accessible to the policy-makers most in need of it, but the ability to communicate was also absent. There are more telephones in Tokyo than in all of Africa[161] and, while it costs negotiators in Madagascar and Côte d'Ivoire $75 to exchange a 40-page text by courier (and it takes five days), the same text can pass (in two minutes) between Canberra and Washington for 20 cents — with copies to every OECD negotiator at no extra time or expense.[162] From the outset then, the bargaining positions of sovereign negotiators are unequal. The inequality is only compounded by the diversity of negotiating fora and the inevitable turf complications that arise. This problem has given rise to Crucible recommendations 7 and 8.

Crucible Recommendation 8

Balancing treaty obligations

The Crucible Group wishes to express its concern over the current confusion surrounding international treaty obligations associated with trade and the environment. Many members of the Group believe that the effective functioning of an international trading system need not conflict with other broad societal goals such as the preservation of the ecosystem. An effective international trading system can support such goals through the creation of a more affluent and cooperative global community. With respect to biological diversity in general and genetic resources in particular, the Group recommends that:

- steps be taken in all relevant fora to address and clarify any actual or perceived lack of compatibility between treaties and to establish monitoring and review mechanisms that will ensure that no disharmony will arise;
- governments ensure that rights and obligations derived from any existing and future international agreement must not cause damage or threat to the long-term security of PGRFA;
- in the completion of a revised International Undertaking on Plant Genetic Resources for Food and Agriculture and the further development of a multilateral system of exchange, governments ensure that the long-term security of genetic resources for food and agriculture is not undermined by implementation of other treaties.

See p. 111 *et seq.* for the Notes

Access and Exchange

The need for — and principle of — access to biological diversity is well understood in history and morality. In the past five years, 'access' has been consistently juxtaposed with 'exchange' — a second, much less practised, need and principle that has sometimes been described as benefit-sharing. Some argue that exchange is not wholly a matter of sharing benefits but of establishing the conditions for exchange and reciprocity between sovereign parties. Two major world fora have become the coaxial centres for a flourishing debate that attempts to ensure the principles and practice of each: FAO, through its Commission on Genetic Resources for Food and Agriculture (CGRFA), and the UN's Convention on Biological Diversity (CBD).

Biological diversity is manifested at the levels of ecosystems, species and genes. Conserving and utilizing biodiversity requires attention to all three levels. The negotiations on access and benefit-sharing are often complicated by the fact that the value we place on different types of diversity is unclear and ever-changing. There is a tendency to prize the diversity we know to have value and to underestimate the diversity we do not recognize as having value. These fluid values often complicate intergovernmental negotiations. The Biodiversity Convention is the legally-binding umbrella for all levels and forms of diversity. Although most people recognize the CBD's centrality, it is nevertheless viewed by some as being functionally more concerned with 'wild' (a contentious term) or 'not-yet-utilized' (no less contentious) diversity. Others see the FAO's Commission as concerned with 'cultivated' or 'nurtured' diversity — that which is known to have 'value'. Such distinctions raise issues of agency territoriality as well as of science. Historically, other divisions arise concerning national sovereignty. Although there is universal recognition that nation states have sovereignty over the biodiversity within their territories, there remain intensely different views on the ownership of biological materials that were removed from a country prior to the coming into force of the Biodiversity Convention. Although the concern spans botanical gardens, zoos, herbariums, tissue culture collections and gene banks, much of the policy attention has focused on *ex situ* collections of agricultural germplasm. This leads policy-makers to another important international forum — the Consultative Group on International Agricultural Research (CGIAR), which holds 40% of the world's unique crop germplasm in storage.

Access and exchange related to major food crops is further complicated because so much of the genetic diversity of the species involved has already been widely dissipated around the world. The centre of origin of a crop is not necessarily the same as its centre of diversity nor, given recent scientific

See p. 111 *et seq.* for the Notes

advances, is the material found within the centre of diversity necessarily as valuable (today) as the crop's weedy relatives or the same species grown under stress conditions in regions far from its homeland.

If FAO, the CBD and CGIAR are in the 'frontline' for negotiation on access and exchange, much of the more specific concern for benefit-sharing reaches out to involve trade and intellectual property legislation and conventions. Some would argue, for example, that the CBD has precedence over the World Trade Organization (WTO) while others would insist that this would be a political absurdity. There is an argument that the South's biodiversity and knowledge have been surrendered 'free of charge' to the North but that the North has claimed both as intellectual property over this material and is now charging the South royalties in order to have access to their own genius. Others argue precisely the opposite: Most of biodiversity represents possible raw material for invention. Intellectual property encourages society to value diversity and ensures that the knowledge gained through 'protected' inventions is shared universally, at least in the long run. Between these two solicitudes range many nuances and variables. The debate surrounding exchange (including benefit-sharing) exposes issues between indigenous and other rural peoples on the one hand and institutional science on the other. It also exposes differences between private and public sectors. Tensions and territoriality between some proponents of intergovernmental institutions such as the WTO and the Union for the Protection of New Varieties of Plants (UPOV) and the CBD, FAO, and CGIAR also arise. The Group's review of outstanding issues logically begins with the Convention on Biological Diversity.

The pivotal position of the Convention on Biological Diversity (CBD)

In December 1993 the CBD came into force, providing an international, legally binding framework for the conservation and sustainable use of biodiversity. The CBD has three main objectives:
- the conservation of biological diversity;
- the sustainable use of its components;
- the fair and equitable sharing of the benefits arising from such use.

The scope of the CBD is enormous, requiring protection of all biodiversity in all types of ecosystems and habitats. The implementation of the CBD has proceeded slowly due to a lack of finance and the difficulty of prioritizing within such an expansive subject area and among 175 Parties with diverse national situations and interests. There is also a level of political uncertainty created by the failure of the US Government to ratify the Convention. Nevertheless, governments have worked reasonably well in establishing important negotiating fora for biosafety, indigenous knowledge, and the analysis of scientific and technical issues. To this end, governments have created a number of instruments including the Conference of the Parties

See p. 111 *et seq.* for the Notes

Policy primer

- Adopted during the Rio Earth Summit, the Convention came into force in December 1993 and currently has 175 member states. (Notably, the USA has yet to ratify the Convention.)
- The CBD is a legally binding framework for the conservation and sustainable use of all biological diversity and is intended to establish processes for the equitable sharing of benefits arising from the use of biodiversity.
- The CBD reaffirms national sovereignty over genetic resources and stresses the importance of *in situ* conservation.
- The CBD is generally interpreted as emphasizing a bilateral approach to access/exchange negotiations between sovereign source countries and recipients.
- The CBD recognizes the central role of indigenous and local communities in biodiversity conservation through their traditional and sustainable practices and knowledge systems.
- The CBD acknowledges intellectual property rights with the understanding that such rights should promote and not compromise the Convention's objectives.
- The CBD is expected not only to oversee and monitor but also to stimulate financial and other resources that will support the conservation and sustainable use of biodiversity.

Outstanding issues:
- Are bilateral and multilateral mechanisms for access and exchange mutually exclusive?
- Are parallel negotiations underway at CBD and at the FAO (International Undertaking) confusing or clarifying access/exchange issues and the role of indigenous and local communities (including Farmers' Rights)?
- Are national obligations under the CBD and the WTO (including TRIPs) compatible?

(COP) itself, the Subsidiary Body on Scientific, Technical and Technological Advice (SBSTTA); the Secretariat, and the Clearing House Mechanism for Scientific and Technical Co-operation (CHM). A financial mechanism operated by the Global Environment Facility (GEF) is mandated to facilitate implementation of the CBD. Much time has been spent (formally and informally) on approaches to access and benefit-sharing.

Bilateral and multilateral mechanisms

The debate over bilateral and multilateral systems of germplasm access and exchange (including benefit-sharing) is central to both the Biodiversity Convention and to the renegotiation of FAO's International Undertaking. While all parties acknowledge that this is not necessarily an either/or discussion, and that there are many hybrid versions that can be considered, the political division also has a scientific basis. Agricultural (crop and livestock) diversity is widely diffused around the globe and is so commonly held in gene banks that bilateral arrangements for the most widely utilized species would seem very problematic. On the other hand, medicinal and other crop species of restricted distribution (e.g. *Hevea*, coffee) are less widely-held and may be more locally confined. Bilateral arrangements may be possible here (though there is debate on this) and may also reasonably identify the sources of knowledge and innovation related to these species.

In Article 15 on access to genetic resources, the CBD sets up a framework of general principles for structuring the international exchange of genetic resources, premised upon the national sovereignty of each country over genetic resources within its jurisdiction, and with the objective of facilitating access to genetic resources rather than imposing restrictions that run counter to the objectives of the Convention. The article, in keeping with the orientation of the treaty generally, focuses on national action and, through reference to mutually agreed terms and prior informed consent, implies a negotiation — a bilateral approach — between source countries and recipients for access to genetic resources. It does not, however, preclude a multilateral approach or system should the Parties choose to adopt such a system for all genetic resources or some subset of these.

National access approaches

Several countries and some regions are developing or have already adopted laws to regulate access to their genetic resources. The Philippines, Thailand and the countries of the Andean Pact (Colombia, Ecuador, Peru, Bolivia and Venezuela) have adopted such measures. In addition, Brazil, Ethiopia, Fiji, Malaysia and India have draft legislation under various stages of consideration. (Volume 2 of this Report discusses various approaches to national legislation for access.)

Most of the laws noted above include similar conditions for access. For example, all or most of the laws require the bio-prospector to:

- submit duplicate samples of any genetic resources collected to a designated institution within the country of origin;
- include a national institution or national researchers in the collection of genetic resources and/or research relating to the genetic resources collected;
- share existing information relating to the genetic resource for which access is requested and any uses of that resource;
- share research results with the competent national authority and the providers of the genetic resource;
- assist in the strengthening of institutional capacities of national institutions relating to genetic resources; and share specific financial or other related benefits (e.g. proprietary technologies).

The laws differ, however, on whether they require a bio-prospector to meet each condition or whether the conditions constitute a group of benefits that the bio-prospector, the provider of the resource, and the State may negotiate. Decision 391 (article 17) of the Andean Pact, for example, allows the parties to negotiate the provisions while Peru's Draft Regulations (article 21) require the bio-prospector to meet each of the conditions of Decision 391 and then a few more. The approach of grouping possible benefits this way has clear advantages for bio-prospectors who are not able to provide the type of information or support that the Peruvian approach requires. With this approach, bio-prospectors and providers can, through negotiation, tailor conditions that are within their means and meet their needs.

See p. 111 *et seq.* for the Notes

Most of these initiatives do not actively discourage conservation, exchange and use of agricultural genetic resources or the creation of a Multilateral System of Access and Benefit Sharing (MUSAB) as under discussion in FAO. However, they do create substantial hurdles, especially for germplasm held by CGIAR institutes. For example, none of the laws clearly and specifically exempt *ex situ* genetic resources acquired before entry into force of the CBD (or the applicable implementing legislation) from access procedures. The laws generally declare that the State has sovereignty over its genetic resources, including *ex situ* genetic resources and derivatives (or derived products), and grants the State the right to determine the conditions of access to those resources. (The complications for FAO and CGIAR are discussed further on.)

Ex situ *collections*
In the absence of other national or international laws or agreements, a research institute hosted in a country with the kind of access rules laid out above could find itself tightly constrained in its use and exchange of *ex situ* germplasm collections even if they originated in other parts of the world. The uncertainty that this creates needs clarification. However, clarification need not imply any curtailment of sovereignty.

This is perhaps the pivotal legal moment for *ex situ* resources. The creation of the CBD represents the first time that the sovereignty of nation states over their own genetic resources was specifically affirmed in an international treaty. Under the FAO International Undertaking on Plant Genetic Resources (IU) of a decade earlier, plant genetic resources were considered to be the heritage of humankind, freely available for use and available without restriction. CGIAR Centre genebanks were widely perceived to be able to collect and exchange germplasm without constraint on the assumption that the flow of crop material benefited all humanity. Following the formation of the CBD, the FAO Commission — an intergovernmental body of over 160 governments — endorsed, in 1994, agreements between the FAO and the CGIAR, declaring designated genetic resources to be held in trust for the international community. But the Commission and the CGIAR have also recognized that these agreements are an interim measure pending the outcome of the current negotiation revising the IU. To whom these resources ultimately belong is therefore difficult to say. Should the IU negotiations fail to conclude an agreement on a multilateral system, there are those who will ask the legal question of whether the CBD's subsequent, different legal regime, which confirms national sovereignty over genetic resources, can change the ownership of resources collected under the previous legal regime. Whether or not the CBD Conference of the Parties can succeed where the FAO Commission failed will be a political question.

The question also arises as to whether a state can nationalize resources acquired as common heritage. This is especially problematic where the genetic

See p. 111 *et seq.* for the Notes

resources in question were collected in another country. The issue is yet more poignant for a country hosting an international institution. If the country is a member of the FAO, it could be assumed that it has acknowledged the principle of common heritage expressed in the original FAO International Undertaking of 1983 and the trust status enshrined in the 1994 Agreements with CGIAR. Some host government agreements also stipulate that genetic resources held by institutes are 'in trust for the benefit of the international community'. Although there remains a great deal of international good will with respect to the exchange of agricultural germplasm, there is no single access law or provision in any country that now specifically addresses the unique and complex issue of genetic resources for food and agriculture. Indeed, governments appear to be drafting legislation or directives with little or no awareness of the implications for agricultural material — material that is arguably the most widely used and vitally needed.

It is difficult to assess the impact of national access legislation when experience is so recent. For example:

- What effect, if any, has the legislation had on the exchange and use of agricultural germplasm? Has there been a drop in the number of requests for access after legislation was introduced? Are transaction costs reasonable and the procedures sufficiently clear and efficient or has the legislation caused parties to look elsewhere for access to genetic resources?
- To what extent have benefits actually accrued from access transactions?
- To whom have benefits actually accrued, e.g. have indigenous and local communities benefited from the transactions?

An access system based on bilateral transactions with the 'country of origin' — the term used and defined in the CBD — is likely to face practical problems as far as agriculture is concerned. Determining the country of origin for most agricultural germplasm will be extremely complex if not impossible. In addition, a bilateral system for agriculture could disadvantage many developing countries and there is no evidence of — indeed, no advocacy for — significant funds flowing back to countries of origin — or that the revenues possible will match the likely transaction costs involved in national or international monitoring of crop germplasm flows.

How can national governments develop access and benefit-sharing legislation for genetic resources that does not undermine or restrict their position in ongoing international negotiations? A Panel of Experts on Access and Benefit-Sharing meeting in Costa Rica in October 1999 offered some clarity and direction on how parties to the CBD could develop national access legislation consistent with existing international obligations, and sensitive to the unique nature of genetic resources for food and agriculture.[163] The Panel agreed on the need for distinct solutions for genetic resources for food and agriculture, such as the development of multilateral regimes. It concluded that Parties, in developing national legislation on access, should

See p. 111 *et seq.* for the Notes

allow for the development of a multilateral system to facilitate access and benefit-sharing for such resources.

Parallel negotiations at the CBD and FAO

Clearly, the preceding discussion emphasizes the importance of close cooperation between the CBD and FAO with respect to access and exchange. Dispassionate observers (if there are any) to the CBD and FAO negotiating processes would all concur that relations between the two sovereign bodies have been 'delicate' but 'improving' in the years since the CBD came into force. The International Undertaking (IU) was negotiated in 1983, ten years prior to the coming into force of the CBD. Not only is the IU not legally binding but it was born at a time of intense South/North conflict in an environment of intense mistrust and little mutual understanding. Further, the focus of the IU has been on agricultural biodiversity. For most of its 15-year career, in fact, the IU has emphasized 'crop' genetic resources and it has only relatively recently come to consider the importance of *in situ* conservation and the vital role of wild and weedy crop relatives, pollinators, and their ecosystem conservation. These issues are discussed in the following section related to the FAO Commission. Policy-makers should be aware, however, that it is folly to consider negotiations at the CBD in isolation from discussions underway at FAO.

Compatibility between the CBD and the WTO
(including intellectual property issues) ·

The negotiations for the creation of the new world trade agreement achieved during the Uruguay Round (1986–93) were concluded in December 1993. The Ministerial Meeting at which it was signed took place in April 1994, in Marrakesh. The agreement came into force in January 1995 — two years after the coming into force of the Convention on Biological Diversity. Given the contentious debate surrounding the creation of each of the treaties, it was inevitable that differences of interpretation would arise that would set the two agreements in conflict. Some of the disagreements could be the result of the deliberate ambiguities built into highly politicized texts as governments sought last-minute consensus. Some of the disagreements are also an extension of the ongoing disputes that arose during negotiations as different factions fight for political space and leverage over issues they feel they lost at the time agreements were adopted. Still other conflicts have arisen as all parties discover that decisions reached late at night have implications no one fully considered at the time. And others have arisen with the evolution of new science and new practices.

The areas of tension between the WTO and the CBD relate to intellectual property and to trade practices that could impact on the conservation of biodiversity through environmental damage. The Crucible Group focused on the perceived conflicts related to intellectual property. Although some of

See p. 111 *et seq.* for the Notes

the discussion is addressed here, policy-makers are advised to follow the discussion later in Volume 2 concerning Innovation, where the WTO TRIPs chapter and important conventions such as UPOV are discussed in much more detail. The key message here is that those national authorities dealing with trade and the environment — as well as agriculture — need to be talking to one another and harmonizing their positions in the relevant fora.

Intellectual property rights are mentioned in Article 16 of the CBD. According to the IUCN/World Conservation Union, Article 16 is an 'ambiguous article whose imprecise text reflects the complexity of the political debate and subsequent compromise reached during the negotiations'.[164] Article 16.2 specifically states that access and transfer of proprietary technology shall be provided on terms that are consistent with the 'adequate and effective protection' of IP.

Article 16 provides for a balance between the protection/recognition of existing IP and the transfer of technology relevant for the conservation and sustainable use of biodiversity, as well as for technology related to genetic resources provided by countries of origin. Article 16.5 is a key provision related to impacts of IPRs on the Convention. It stipulates that IPRs should promote and not run counter to the objectives of the Convention (a commitment seen as eminently achievable by some, and as an oxymoron by others).

> *Article 16.2* Access to and transfer of technology referred to in paragraph 1 above to developing countries shall be provided and/or facilitated under fair and most favourable terms, including on concessional and preferential terms where mutually agreed, and, where necessary, in accordance with the financial mechanism established by Articles 20 and 21. In the case of technology subject to patents and other intellectual property rights, such access and transfer shall be provided on terms which recognize and are consistent with the adequate and effective protection of intellectual property rights.

> *Article 16.5.* The Contracting Parties, recognizing that patents and other intellectual property rights may have an influence on the implementation of this Convention, shall cooperate in this regard subject to national legislation and international law in order to ensure that such rights are supportive of and do not run counter to its objectives.

IPRs have been discussed at various meetings of the Conference of the Parties (COP) and its subsidiary body, though no decision has been reached about their impacts on the objectives of the Convention. With respect to access to genetic resources and benefit-sharing, some Parties believe that existing IPRs could be instrumental in benefit-sharing mechanisms because they could allow users of genetic resources to generate revenues from their

See p. 111 *et seq.* for the Notes

Viewpoint box: Are the CBD and the WTO TRIPs chapter compatible?

Leave it to national governments

The two Conventions have been written carefully and, whilst dealing with different matters, both contain provisions related to the other one. CBD considers that intellectual property can be made supportive of its objectives. TRIPs excludes from patentability inventions that are contrary to the *ordre public* or morality, or those inventions that are dangerous to animal or plant life or seriously prejudicial to the environment. These are important safeguards.

Parties to both agreements must fulfill their obligations under both treaties and through thoughtful implementations there is no reason why they cannot do so without jeopardizing the objectives of either.

TRIPs is in conflict with CBD

TRIPs obligates members to adopt patents or *sui generis* systems for plant varieties, while CBD calls for the protection and promotion of indigenous knowledge, innovations and practices. Conservation and privatization are contradictory goals. Exclusive monopoly rights over biological products and processes restrict availability of genetic resources — and this is detrimental to food security and the wellbeing of local communities. Western-style IP regimes will promote uniformity and the introduction of new plant varieties that unintentionally displace farmers' varieties. Patents are not benefit-sharing agreements, but are means by which the private sector, mostly in the industrialized world, is able to profit from the biodiversity that was conserved and developed by indigenous and local communities over millennia. The rights and objectives of the two treaties are clearly in conflict. Both treaties provide legally binding obligations for governments; which takes precedence?

TRIPs supports CBD

The slogan 'patenting of life' is emotionally effective but not accurate. According to the TRIPs Agreement, microorganisms which are living matter, must be patented. But TRIPs does not address the patenting of life *per se*, nor does it make it necessary for countries to patent higher life forms. That being said, TRIPs does require countries to implement effective *sui generis* protection (which is not as exclusive in nature as patents) for plant varieties. There is no inconsistency between private intellectual property rights granted for a well-defined invention, for a limited period of time, and the sovereign rights of nations over their biological resources.

If we want the private sector to be interested in developing new drugs, or new plant varieties from wild biota, it must have the possibility to protect the results of its work. Protection is the best trigger point for the sharing of benefits. Protection laws do not deprive local communities of continued use of their indigenous products and processes. The requirements for obtaining a patent are novelty, inventive step (non-obviousness) and industrial applicability (usefulness), and granting offices rigorously respect these criteria. Furthermore, indigenous knowledge may be the foundation on which a novel patentable process or product is developed. When this happens, industry believes that this must be acknowledged by the inventor and compensation should be provided for on mutually agreed terms, as required by CBD.

Modern plant breeding is enhanced and promoted by IP legislation. The development of new plant varieties for food and agriculture increases biodiversity at the disposal of farmers. We can't measure crop genetic diversity merely by a count of farmers' varieties — as they are often very similar in terms of genetic background.

See p. 111 *et seq.* for the Notes

inventions, which they could share with countries of origin or local communities. Others believe that existing IPRs do not serve this purpose, that they are inadequate to protect the rights of farmers and indigenous peoples, and that these laws constitute one of the greatest threats to the future conservation and enhancement of biodiversity. IPRs, they assert, do not promote benefit sharing.

Given the diversity of views on the subject, the third meeting of the COP called for case studies on the impact of IPRs on the objectives of the Convention, which are still under review. Accordingly, the COP has come to pay special attention to discussions in other fora concerning intellectual property. The Secretariat has asked for and been granted observer status at the Committee on Trade and Environment of WTO and is seeking observer status at the WTO's TRIPs Council.

The agricultural biodiversity mission of the FAO Commission on Genetic Resources for Food and Agriculture (CGRFA)

Policy primer

- CGRFA began in 1983 and held its first meeting in 1985. It normally meets every two years but has held a number of extraordinary sessions related to both Leipzig and the revision of the International Undertaking.
- 164 countries and the European Community are members of the Commission or the IU or both.
- Aside from responsibility for the IU, the Commission also oversees the FAO-CGIAR Trust Agreement and provides policy oversight for the germplasm collections included in that agreement.
- In 1995 the Commission extended its scope beyond plant genetic resources to encompass livestock. It will also review similar work related to forests and fisheries as these areas relate to food and agriculture;
- Through the Commission, governments monitor the follow-up to the Leipzig Global Plan of Action.

Outstanding issues:
- Are governments willing to strengthen FAO's role in genetic resources for food and agriculture?
- Are governments prepared to adopt a legally binding multilateral agreement for access and exchange of agricultural plant genetic resources (the International Undertaking)?
- Are Farmers' Rights a matter of national implementation or international human rights? (See discussion under section on 'Knowledge'.)
- Can governments achieve an intelligent balance in functions between the CBD and the FAO Commission?
- Will governments fully implement the Leipzig Plan of Action?

Strengthening FAO's Global System for the Conservation and Sustainable Use of Plant Genetic Resources

Since 1983, member nations of the United Nations Food and Agriculture Organization have taken important steps to resolve these contentious

See p. 111 *et seq.* for the Notes

questions by establishing a Global System for the Conservation and Utilization of Plant Genetic Resources for Food and Agriculture (PGRFA). The aims of the Global System are:

- conservation of biological diversity;
- sustainable use of its components;
- fair and equitable sharing of the benefits arising from the utilization of genetic resources.

The Commission on Genetic Resources for Food and Agriculture monitors the development of the Global System. The main institutional component of the Global System is the International Undertaking on Plant Genetic Resources. The Commission was founded in 1983 and held its first meeting in 1985. The Commission provides an intergovernmental forum where countries — as donors and users of germplasm, funds and technologies — can meet, on an equal footing, to discuss and reach consensus on matters related to crop germplasm. Its mandate (and name) was broadened to include all genetic resources for food and agriculture in 1995. Today, 160 countries and the European Community are members of the Commission.

As delegates to any UN meeting can vouch, each of the specialized agencies can, at the drop of a hat, produce charts, graphs and management designs that will clearly prove the central position in the universe occupied by their work. Although FAO is far from immune to these temptations, there is no disputing that FAO has played a major and a pioneering role in drawing the world's attention to the urgent need to conserve and utilize agricultural germplasm. The Commission's secretariat, in particular, has served the international community well and brilliantly. Nevertheless, some Crucible members are concerned that the full weight of FAO's technical expertise is not behind the work of the Commission and that the so-called 'System' — including databases and gene bank networks, championed in reports and documents — is more of a paper tiger than a practical reality. Governments must make it possible for FAO to carry out the mandate with which it has been charged and FAO must make clear to governments where the romance ends and the reality begins. The Commission is the political forum for agricultural germplasm debates. The International Undertaking is the basis for negotiation and its revision is critical to the issues of access and exchange.

The revision of the International Undertaking
In the late 1970s, developing countries and CSOs first raised political concerns over control, ownership and access to plant genetic resources for food and agriculture (PGRFA) at the FAO. At that time, PGRFA were considered by virtually all governments to be the 'common heritage of mankind'. Developing countries at FAO began to argue, however, that common heritage implied the existence of common responsibilities to safeguard genetic resources. They also began to question the contradictory nature of free access

See p. 111 *et seq.* for the Notes

Policy primer

- Resolution 3 of the Nairobi Final Act in 1992 recognized that access to *ex situ* collections not acquired in accordance with the CBD and Farmers' Rights are outstanding issues for which solutions should be sought within the FAO Global System.
- The FAO International Undertaking on Plant Genetic Resources is a non-binding intergovernmental instrument adopted in 1983. 113 countries are signatories to the IU. It is intended to facilitate access, conservation and sustainable utilization of plant genetic resources.
- The International Undertaking is in the process of being revised in harmony with the CBD. Three outstanding issues dominate the current IU negotiations: scope and access, benefit sharing and Farmers' Rights — issues with significant implications for the CBD and for intellectual property regimes.

to plant genetic resources of the developing world in the face of proprietary rights for new plant varieties developed by institutional plant breeders. They asked: Why are proprietary seeds, originating in the developing world, bringing royalties to institutional plant breeders in the industrialized world without corresponding compensation for the original donors/innovators of the genetic material? Who is responsible for conserving plant genetic resources? Who controls access to genetic material, and what mechanisms are needed to ensure reciprocal benefits between the industrialized world and the developing world?

It was with this political background that governments created the International Undertaking in 1983. Now, in the light of the Rio Earth Summit and the CBD, governments are working to update the IU to address new issues — or the same old issues dressed for the nineties.

In the years following its adoption, many governments came to recognize that the IU is incomplete and contains ambiguities requiring clarification. Over the past decade, in fact, three interpretative resolutions were adopted to clarify concepts and terms in the IU.[165] One of the main objectives of the current revision of the International Undertaking is to harmonize its provisions with the CBD's principles of national sovereignty over PGRFA, access, prior informed consent and sharing of benefits. The CBD negotiators specifically requested that FAO's Global System resolve two outstanding issues of particular relevance to agricultural biodiversity — the question of Farmers' Rights and the status of *ex situ* collection in existence prior to the CBD. Negotiations to revise the IU were initiated in this context in the mid-1990s, although it has not yet been decided, in its final form, whether the IU will be (1) under the FAO constitution, (2) on its own as a stand-alone treaty, or (3) under the auspices of the CBD as a protocol.

Renegotiation of the International Undertaking was on the agenda of the April 1999 session of the Commission on PGRFA. A single negotiating text was accepted as the basis for negotiation.[166] There is a basic agreement to develop a multilateral system of access to PGRFA. Under such a system, access to PGRFA and the resulting benefits would be open to all parties who

See p. 111 *et seq.* for the Notes

are participants in the multilateral system on access and benefit sharing. Three outstanding issues dominate the current IU negotiations: scope and access, benefit sharing and Farmers' Rights.

Scope and access
The scope of what plant genetic resources the IU will cover, and the question of how access will be regulated, are key issues in the negotiations. It is generally agreed that the scope of the revised IU would cover all plant genetic resources for food and agriculture, but that governments may wish to determine a multi-tiered system of access ranging from 'free access' to certain species under a multilateral system, to a 'not free' or restricted access system subject to bilateral exchange agreements (which could be monitored or maintained with the assistance of a multilateral facility). There is a growing willingness to accept a relatively free flow of germplasm among signatory states for some food crops already widely dispersed and utilized by plant breeders. However, plant species less widely available and/or having high commercial value (i.e. coffee, high-value spices or medicinal plants) would be restricted and their access controlled, subject to bilateral negotiations. Initially, a limited range of crops would be available for multilateral exchange, but the lists would be maintained as an annex to the IU, and could be dynamic over time. However, some countries pointed out during the April 1999 negotiations that ongoing revision of the lists is problematic.

Many issues are not finalized. Will the revised IU apply to materials collected both before and after the CBD entered into force, or just those acquired prior to the CBD? Will CGIAR collections have a separate status? Will non-parties to the IU be denied or restricted in their access to materials (including those of the CGIAR)?

Generally speaking, many developing countries approach 'scope and access' defensively. There is concern that they are being asked to donate their germplasm freely while some industrialized countries allow for claims to temporary exclusive monopoly on the same material through intellectual property regimes. A number of independent studies have suggested that — while collectively invaluable — it is next to impossible to determine the commercial worth of *single* germplasm accessions. In fact, the monitoring costs of tracking gene flows could exceed the actual financial returns to countries of origin.

The status of germplasm protected under intellectual property is controversial. For many developing countries, it is particularly important that this germplasm should be fully part of the 'free' multilateral exchange system. This is material with defined commercial value and is a vital part of the contribution from the North, necessary to balance the South's richer store of unexploited potential. IP owners, however, contend that IP-protected material is already freely available, to the extent appropriate — which (they say) is for use in breeding development, rather than in direct commercial

See p. 111 *et seq.* for the Notes

exploitation of unmodified material. Material protected by plant variety rights can be freely used in further breeding. The research exemption for patented material is generally more restrictive. In the United States it appears that the research exemption is so narrow that researchers are not allowed to use propagating material that includes patented material within it, even if the patented material is bred out of (i.e. not included in) the researcher's final product. The law in the UK on this point appears to be somewhat more lenient; there it seems that the researcher can use the material to create something new, as long as the patented portion is not in the final product. In jurisdictions that have practically useful research exemptions, the only material that researchers could not use would be that which is protected by trade secret law. Exemptions from patent holders' rights are analysed in more detail in Volume 2, Topic 3, Section 3, 'Options for intellectual property laws for biotechnological inventions'.

CSOs offer another range of views on the issue of access to PGRFA. Some believe that farming communities should either develop their own intellectual property regimes for farmers' varieties — or adopt non-IP, *sui generis* mechanisms to defend their rights. Others regard any non-traditional constraints to germplasm exchange (be they IP or non-IP, *sui generis*) as self-defeating since such approaches, they argue, would undermine community plant breeding.

Benefit-sharing

How will countries that are donors of germplasm to the multilateral system share in the benefits derived from those materials? What is the benefit-sharing mechanism? Should it be voluntary or mandatory? Some governments argue that developing (and other) nations are adequately compensated for their germplasm merely by having access (free or royalty-tied) to the information and breeding material created by open, international exchange. Others consider this a 'trickle-down' approach that does not adequately recognize the contribution of farming communities and national governments. Efforts to proportion the contribution and benefits from germplasm exchange and plant breeding have left all sides uncertain and frustrated, especially in light of the fact that there are few funds available, even in industrialized countries, to support this goal. National budgets for PGRFA are facing severe constraints and cutbacks, in both industrialized and developing countries.

Some parties suggest that benefit sharing might best be achieved through the full implementation (including new and additional financing) of the Leipzig Global Plan of Action (GPA). Various scenarios have been devised under which industrialized countries would make financial contributions to the GPA through membership fees to the multilateral exchange 'club', while developing countries (as well as industrialized countries) would make their germplasm available to club members. In effect, the membership 'fee' for

See p. 111 *et seq.* for the Notes

developing-country members would be germplasm rather than money. Again, conditions for scope and access might vary and one or more species categories could be entertained. All members of the multilateral system would be able to access the germplasm and the funds on the basis of approved programs and projects in line with the rolling GPA. Not surprisingly, there are many nuances to this scenario and many concerns about prohibiting IP claims on germplasm in the multilateral system.

There is increasing recognition that the future role of the CGIAR's international centres, and its mandate to conduct agricultural research in service to the world's poor, is one possible concrete mechanism for sharing the benefits of PGRFA. In light of the growing importance of the CGIAR centres, there is concern among governments and CSOs that aspects of ownership, trusteeship, governance and the rules regarding access and exchange of CG germplasm be strengthened and clarified.

At the First Inter-sessional Meeting of the Chairperson's Contact Group of the Commission on Genetic Resources for Food and Agriculture, held in Rome, 20–24 September 1999, negotiations focused on benefit-sharing within the multilateral system. All non-governmental observers were excluded from participation, with the exception of IPGRI, representing the CGIAR. There was general agreement that some global benefits could arise from the use of PGRFA under a multilateral system, through the exchange of information, access to and transfer of technology, capacity-building and the sharing of benefits on commercialization. Many outstanding issues are still under negotiation, including financial matters. For example: while access to technologies, improved varieties and germplasm will be provided and facilitated, will it be subject to applicable property rights and access laws? Are there special circumstances under which access to and utilization of information and technology protected by IP agreements and confidentiality can be made freely available? How can parties give practical expression to the fair and equitable sharing of commercial benefits? How will parties establish and/or strengthen programs on capacity-building, and who will determine priorities? What role will the Global Plan of Action play? Negotiations on these issues will continue at the next meeting of the Chairman's Contact Group in early 2000.

The issue of Farmers' Rights

The principle of Farmers' Rights, endorsed by the FAO in 1989, recognizes that farmers and rural communities have contributed greatly to the creation, conservation, exchange and knowledge of genetic resources, and that they should be recognized and rewarded for their past and ongoing contributions.[167]

Many governments and CSOs have embraced the principle of Farmers' Rights, not only as a counterpoint to Plant Breeders' Rights (as it was initially proposed), but also as recognition of the innovative role that farmers and

See p. 111 *et seq.* for the Notes

rural communities play in the conservation and further development of genetic resources and their right to benefit from it. In 1991, the Keystone Dialogue (held by a multi-stakeholder group that brought together representatives from government, CSOs, scientists and the private sector) attempted to elaborate the meaning of Farmers' Rights by suggesting that it includes the right to germplasm, information, financial resources, technologies, and systems of research and marketing. The Keystone Group suggested that a concrete way of recognizing Farmers' Rights would be to create a fund to support genetic conservation and utilization programs. It was also agreed in Keystone that farming communities had the right to save and exchange seed. (Although an international fund was created in the 1980s by the FAO Commission as a channel for countries, governments, CSOs, private industry and individuals to support conservation and sustainable use of plant genetic resources, it was voluntary and existed primarily on paper.)

Perhaps because Farmers' Rights was conceived as a political term, it has been especially difficult to define it in legal terms. It has been debated continuously over the past decade, and is a key aspect of the ongoing renegotiation of the International Undertaking.

Support for the 'principle' of Farmers' Rights is a mile wide and (some would argue) an inch deep. As the IU negotiations proceed at FAO it remains unclear whether or not Farmers' Rights will ever be implemented — or whether their implementation will be solely at the national level and not at the international level. CSOs and some governments have posed additional elements for Farmers' Rights that may go beyond its original definition. Critics insist that the range of claims being made under the Farmers' Rights umbrella far exceed the mission of either the FAO Commission on Genetic Resources for Food and Agriculture (CGRFA) or the CBD and must either be reduced or dropped altogether. Others wish to limit the debate on Farmers' Rights within CGRFA to the right to save seed and to have access to genetic resources while leaving wider land and culture issues to the UN High Commissioner for Human Rights. Still others see Farmers' Rights as a vague statement of principle and appreciation but nothing more and insist that the realization of Farmers' Rights can only occur at the national level.

In the context of PGRFA, implementation of the Leipzig GPA (see next section) would be one way to implement Farmers' Rights effectively. Some have suggested that the implementation of Farmers' Rights through the Leipzig GPA could mean that farming communities might have special access to resources and special recognition in the governance structure for the multilateral exchange system or 'club'. Following the World Food Summit, others have suggested that some elements of Farmers' Rights might be included in the current review of 'the Right to Food' being jointly undertaken by the UN Human Rights Commission (UNHRC) and FAO. A study on the

Right to Food submitted to the Human Rights Commission urges that Farmers' Rights should be addressed by the human rights community and promoted as part of the right to food.[168] A third element has evolved along the lines of the right of farming communities to 'say no' — to opt out of germplasn. exchange or enhancement systems they regard as counterproductive.

At the April 1999 Commission meeting, the debate on Farmers' Rights focused on the right of farmers to save seed from their harvest. Ten years before, when the FAO Commission recognized that Plant Breeders' Rights were not incompatible with the IU's objectives, the prevailing model of PBR (the 1978 UPOV Act) was widely presumed to permit farmers to re-use proprietary seeds. Since then, however, there has been a general tendency in government and industry to constrain or curtail the unlicensed replanting of proprietary material through national intellectual property legislation or through new scientific developments such as genetic seed sterilization (genetic use restriction technology or 'terminator' technology). Although the situation is hotly disputed, UPOV's 1991 Convention has also been widely interpreted as a move to prevent farmer plant-back. The UPOV 1991 Act contains an option for a farmers' exemption under which UPOV member states may exempt farm-saved seed from the breeders' right 'within reasonable limits and subject to safeguarding the legitimate interests' of the breeder. While the 1978 UPOV Act allowed implicitly for farmers to exchange limited amounts 'over the fence' the 1991 UPOV Act allows for farmers to use for propagating purposes on their own holdings only those seeds that they have obtained from their own holdings. This casts doubt on the possibility of exchanging seeds for planting on one's own holding. However, UPOV provides that subsistence farmers (and others) may use protected seed 'for purposes that are private and non-commercial'. These issues became central to the debate at FAO.

Many members of the Crucible Group agree, but for different reasons, that the article on Farmers' Rights adopted by the Contact Group in April 1999 was contentious and is likely to be opposed by some governments and CSOs in future negotiations. The agreed text concludes that responsibility for realizing Farmers' Rights, as they relate to PGRFA, rests with national governments.[169] Article 15.3 states: 'Nothing in this Article shall be interpreted to limit any rights that farmers have to save, use, exchange and sell farm-saved seed/propagating material, subject to national law and as appropriate.' CSOs warn that the current text establishes the primacy of national patent laws over Farmers' Rights. The concern is that the agreed text would allow national governments to use intellectual property laws to prevent farmers from saving and exchanging seed. Meanwhile, some plant breeders and governments warn that the agreed text does not expressly exclude plant varieties protected either by patents or Plant Breeders' Rights.

See p. 111 *et seq.* for the Notes

The Contact Group's decision in April 1999 will be fought and refought for years to come as governments and observers anguish over their choices and trade-offs. Here are three viewpoints on this issue.

Viewpoint box: Did the FAO Commission satisfactorily resolve the issue of farmer plant-back?

No
For the first time, the South abandoned the possibility of establishing Farmers' Rights as an element of Human Rights and the Right to Food. By the tying of the right to save seed to national laws, the Human Rights of farmers have been taken away. Worse still, governments surrendered Farmers' Rights largely by accident when they rushed a late-night decision they should have slept on.

It's not over yet
Farmers' Rights will come back on the table for further negotiation toward the end of the IU revision as governments weigh the plusses and minuses of the whole accord. Then, or sooner, governments can propose that elements of Farmers' Rights be regarded as Human Rights.

Yes
For the first time, the right of farmers to save, use, exchange and sell farm-saved seed is affirmed in the strongest possible terms in an international agreement. Of course, this right is subject to national law since only national law can protect farmers within a country, in accordance to national priorities and needs. The FAO Commission doesn't even have the mandate to discuss and/or negotiate Farmers' Rights as an element of Human Rights.

The next meeting of the Chairman's Contact Group is planned for early 2000. All observers from civil society organizations are excluded from participation in the Chair's Contact Group. An agreed text is likely to be debated at an extraordinary session of the Commission no later than July 2000. If approved by the Commission on PGRFA, the IU will go to the FAO Council in November 2000. (The issue of Farmers' Rights appears again later in this report in the context of other negotiating fora.)

Crucible Recommendation 9

Finalizing a new International Undertaking (IU)
Recognizing that the completion of the negotiations concerning the new IU shall and must be completed in the very near future, the Group recommends that:
- the IU be legally-binding upon signatory states;
- the IU be presented to the Conference of the Parties of the CBD for its consideration as a possible protocol to the Convention with the understanding that the IU be governed through its own Conference of the Parties and operate with an independent secretariat perhaps administered by the FAO.

The Convention on Biological Diversity
In spite of the CBD's sweeping mandate to protect biological diversity, there are two outstanding and particularly problematic areas related to agricultural biodiversity that were identified before governments signed the CBD in Rio de Janeiro.[170] The CBD negotiators passed a resolution

See p. 111 *et seq.* for the Notes

requesting that FAO resolve the questions of: 1) Farmers' Rights (see discussion above); and, 2) the status of *ex situ* collections collected prior to the CBD entering into force. The FAO Commission subsequently initiated negotiations to revise the IU in harmony with the CBD and to address these outstanding issues. (See separate discussion under International Undertaking above.)

The CBD's decision to cede the issue of Farmers' Rights and *ex situ* collections to FAO underscores some of the complexities and peculiarities of dealing with agricultural genetic resources in a bilateral framework. In contrast to those genetic resources that are rare and geographically localized, such as wild plant species of pharmaceutical interest, the market value of agricultural biodiversity is not easily established, and it is especially difficult to determine its origin because of the widespread diffusion and adaptation of crop genetic resources worldwide. A bilateral access and benefit-sharing framework for agricultural biodiversity is, in most cases, extremely complex, if not impossible. Negotiations underway on the revised IU are considering a broader, multilateral approach to access and benefit-sharing for some major crops which seeks to avoid the problem of linking benefits or compensation to specific transactions for specific genetic resources, and tying those benefits to a single country of origin.

In the light of the requests made by governments to FAO, the FAO-CGIAR Trust Agreements related to agricultural germplasm and the efforts of the Commission to revise the IU and implement Farmers' Rights — along with FAO's leadership in establishing the Leipzig Global Plan of Action — are all significant contributions to achieving the goals of the CBD. Nevertheless, as already noted, tensions have arisen between the two organizations that have frustrated cooperation. While the situation is recently much improved, there remains some uncertainty as to whether or not the revised IU should be adopted solely by sovereign states within the framework of FAO or if the revised IU should (or must) also be accepted by the Conference of the Parties of the CBD.

The Leipzig Global Plan of Action
In the early 1990s, FAO spearheaded an international, country-driven process designed to ask questions about the state of the world's agricultural diversity, and to identify the actions needed to insure that it is conserved, utilized and further developed. The four-year preparatory process drew on the active participation of all major actors in the bio-policy and conservation arena — including 158 national governments, scientific institutions, the private sector, CSOs, farmers' organizations and other community-based conservation experts. More than 2000 specific recommendations were tabled during national and regional preparatory meetings.

This process culminated in June 1996 when high-ranking officials from ministries of agriculture, foreign affairs and the environment of some 150

See p. 111 *et seq.* for the Notes

Policy primer

- Following an exhaustive four year country-driven process that led to 150 national plans of action and more than 2000 recommendations, governments in 1996 in Leipzig adopted a Global Plan of Action (GPA) for the conservation and utilization of plant genetic resources.
- The Leipzig GPA contains 20 priority programs emphasizing the rationalization of *ex situ* germplasm collections and the development of *in situ* conservation strategies along with a strong orientation toward sustainable utilization.
- There was no agreement on financing for the Plan (generally estimated at $130-300 million per annum) and the Leipzig Plan has not been formally operationalized at the international level.
- Responsibility for the implementation of Leipzig rests with governments through the FAO Commission (CGRFA).

countries met in Leipzig, Germany, for FAO's Fourth International Technical Conference on Plant Genetic Resources for Food and Agriculture. The Leipzig Conference adopted the first-ever Global Plan of Action, costed at approximately $131–304 million per annum (1997–2007), although the costing was not directly accepted by the meeting.

The Leipzig Conference also considered the 'FAO Report on the State of the World's Plant Genetic Resources', the first comprehensive assessment of the status of plant genetic resources and existing capacity to conserve and utilize them.

Despite the existence of a variety of sources of financing for the conservation and sustainable use of Plant Genetic Resources for Food and

Crucible Recommendation 10

Implementing a Global Plan of Action

The Crucible Group considers the development and adoption of the 1996 Leipzig Global Plan of Action to be a major achievement of the international genetic resources community in the last decade.

The Group expresses its deep disappointment that neither FAO nor the CGRFA have followed through effectively on the implementation of the GPA.

Bearing in mind the accomplishment of the Leipzig process in establishing a rolling GPA, the Group recommends that:

- FAO reinvigorate the country-driven process for the implementation of the GPA that originally led to the adoption of the GPA in Leipzig;
- a fully funded GPA be acknowledged by signatory states to the IU as one possible and important mechanism, among others, for equitable benefit-sharing within a multilateral system of access and exchange;
- the Global Forum on Agricultural Research, at the time of its meeting in Dresden in 2000, be asked to undertake an evaluation of impediments to progress since the adoption of the GPA, including an analysis of the institutional and financial arrangements and contributions.

See p. 111 *et seq.* for the Notes

Agriculture (PGRFA), there are gaps, overlaps, inefficiencies and unnecessary redundancies in the activities financed. The GPA aims to focus resources on the priorities that have been identified at the various levels, and increase the overall effectiveness of global efforts by providing linkages and coordination. The GPA has 20 priority activities in the following categories: 1) *In situ* conservation and development; 2) *ex situ* conservation; 3) utilization of plant genetic resources; 4) institutions and capacity-building.[171]

At the outset of the process leading to the 1996 Leipzig Conference, it was hoped that the evolution of the Global Plan of Action would be paralleled by the revision of the International Undertaking (formalizing an agreement on access/exchange) and that both the GPA and IU would be finalized simultaneously. This proved impossible. The GPA was separated from the issue of Farmers' Rights. It was also separated from the issue of access. Largely for this reason, the GPA has not been fully implemented at the international level. Although some countries and the CGIAR have begun to reorient their work to support the goals of the GPA, it still lacks dedicated funding and an agreed mechanism for monitoring, implementation and support.

The unique role of the CGIAR in facilitating access and exchange

Policy primer

- CGIAR oversees a network of 16 international agricultural research centres (IARCs), the largest agricultural research effort in the developing world with an annual budget in excess of $340 million.
- Collectively, research from the 16 IARCs provides enhanced germplasm that helps feed at least 2 billion people every day.
- CGIAR manages approximately 600 000 agricultural seed samples which amount to approximately 40% of the world's unique germplasm in storage worldwide.
- The CGIAR is supported by more than 40 governments, foundations, and research institutes through informal meetings and collegial understandings. Despite its influence, the CGIAR has no collective legal identity.
- FAO and the CGIAR centres with germplasm collections have signed agreements bringing the collections under the auspices of FAO. The FAO Commission on PGRFA determines the policies under which the network operates.

Outstanding issues:

- Has CGIAR solved the problem of international *ex situ* germplasm collections?
- What role could the CGIAR play in a long-term approach to access and exchange?
- Does the development of intellectual property with respect to plant germplasm have any implications for CGIAR research and germplasm management?

The Consultative Group on International Agricultural Research, established in 1971, is an informal association of public and private donors that supports a network of 16 international agricultural research centres (IARCs), each of which has its own governing body. The CGIAR's mission is to use science

See p. 111 *et seq.* for the Notes

and technology, in partnership with other organizations, to increase food security, alleviate poverty and protect the environment. With a budget of approximately US $340 million per annum, the CGIAR oversees the largest agricultural research effort in the developing world. The CGIAR Secretariat is housed in the World Bank (Washington, DC) and the group's major donors include the World Bank, Japan, the USA and the European Union.

Through its 16 international institutes, the CGIAR manages approximately 600 000 seed samples which amount to approximately 40% of the world's unique germplasm in storage worldwide. Because IARC gene banks contain 'inventoried' germplasm, their collections are considered among the most valuable genetic materials, both because they are more readily identifiable and accessible to institutional breeders than farmers' varieties or wild crop relatives and because of the large amount of information available on the individual samples. The vast majority of crop germplasm held in the IARCs was collected from farming communities in the developing world. But to whom that treasure ultimately belongs, and to whom a genebank is accountable, has been the subject of controversy and debate.

The issue of internationally-held *ex situ* collections (and the 1994 FAO/CGIAR Trust Agreements)

As an interim step in addressing the status of *ex situ* collections, the CGIAR and FAO signed agreements in October 1994 that place most of the gene bank material from CGIAR centres under the auspices of FAO, to be held 'in trust' for the world community. Under the 1994 Trust Agreement, all designated germplasm is to remain in the public domain, ensuring as far as possible a continuing unrestricted flow of germplasm to all researchers. The FAO Commission on Genetic Resources for Food and Agriculture is responsible for setting policy for a network of *ex situ* collections, which includes the centres' in-trust collections.

CGIAR centres routinely distribute germplasm to plant breeders through material transfer agreements (MTAs). The Trust Agreement also requires CGIAR to prohibit recipients of designated germplasm from applying for IPRs on the germplasm or related information. (Recipients who have not signed MTAs — for example those who obtained material prior to 1994 — are not legally bound by these restrictions.) The Trust Agreement does not prohibit CGIAR patenting or claiming any other IPRs with regard to material isolated from designated material, such as cells, genes, etc. The CGIAR guidelines, which have never been formally adopted but only 'accepted', require the authorization of the centre where a recipient wishes to apply for IPRs with regard to such isolated materials. However, these guidelines do not form part of the FAO/CGIAR Trust Agreement. Furthermore, the MTAs allow breeders to use designated material for breeding purposes and to apply for Plant Breeders' Rights with respect to their new varieties.

In late 1997 and 1998, documented examples came to light wherein

See p. 111 *et seq.* for the Notes

recipients of CGIAR designated germplasm applied for intellectual property or otherwise asserted proprietorship over germplasm. While neither governments nor breeders are obliged to adhere to the FAO–CGIAR Trust Agreements, these acts were in contradiction to those agreements as endorsed by the governments of the institutes involved. Of course, insofar as governments or breeders are parties to the contractual provisions enshrined in a material transfer agreement they are obligated to respect those terms. These examples raised serious questions about FAO's and CGIAR's capacities to implement the accords and whether or not the trust arrangement served a useful purpose. Special concern was raised with respect to material transfer agreements used by CGIAR centres when exchanging trust germplasm.

In response to the situation, the Chair of CGIAR called for a moratorium on the granting of IP rights on all designated plant germplasm.[172] In October 1998, FAO and CGIAR adopted a new procedure confirming what each party would do if trust germplasm becomes the subject of proprietary claims.

According to some CSOs, abuses are widespread. In September 1998 the Canadian-based Rural Advancement Foundation International (RAFI) and the Heritage Seed Curators Australia (HSCA) released a report documenting 147 examples of plant intellectual property claims (PBRs and patents), which they considered dubious.[173] According to the HSCA/RAFI report, in more than one-third of the cases cited the plant varieties were collected in foreign countries and submitted for PBR without any evidence of breeding.[174] According to RAFI and HSCA at least 16 of the suspect claims — granted and/or pending — are believed to involve germplasm under the FAO Trust Agreement.[175] The report charges that systematic abuses are taking place showing that plant patents are predatory on breeding work undertaken by farmers and indigenous people around the world.

Breeders' associations and UPOV, while admitting that rights may sometimes be granted when not justified, deny that this is a general problem. They do not accept that any abuse has been proved. Registration procedures may not call for proof of breeding work — in any case, absence of evidence is not evidence of absence. Even if all the alleged 147 abuses were conclusively proved, argues the International Association of Plant Breeders for the Protection of Plant Varieties (ASSINSEL), these represent a mere 0.45% of all Plant Breeder's Rights granted over the past five years, and only 0.15% of the samples distributed by the CG centres. ASSINSEL points out that most of the alleged violations involve public breeders, who are often not completely aware of the complex system of IP and MTAs. It would not be cost-effective, argues ASSINSEL, to spend a great deal of money to monitor so few violations. In response, RAFI and HSCA note that their study focused primarily on 118 Australian claims which amount to 6% of all applications made to the Australian PBR office since the legislation was adopted in that country. The PBR violations, asserts RAFI, were those that were clearly visible based on available documentation — a more in-depth and well-

See p. 111 *et seq.* for the Notes

funded study would likely uncover many more violations. ASSINSEL, however, believes that such a study would show that many of the original allegations were unfounded.

Ultimately, six Australian PBR claims were abandoned or withdrawn in 1998 after alleged abuses were brought to light.[176] In addition, the Australian PBR Office in Canberra has published revised standards for assessing PBR applications that are designed to improve protocols and correct potential abuses.[177] Nevertheless, in May of 1999, RAFI uncovered yet another instance, also in Australia, where a government institute was apparently claiming rights for a CGIAR-bred wheat variety.[178] Prompt action by the centre involved put an end to the claim.

What is designated germplasm under the FAO/CGIAR Trust Agreement? The Trust Agreement states that CGIAR centres shall not claim legal ownership over designated germplasm or related information. According to guidelines elaborated by the International Potato Centre (CIP), 'related information' refers to passport and characterization data, and when available in the databases of the respective genebanks, evaluation data and information on indigenous knowledge.[179] Cells, organelles, genes, etc. from designated germplasm are not specifically mentioned in the agreement. The general rule of law is that recipients of materials do not require the permission of the provider to patent further innovations derived from such materials. The CGIAR Guiding Principles adopted in October 1996 require centres to impose obligations on recipients that they protect such derived inventions only with the agreement of the supplying centre. However, the Guiding Principles do not form part of the FAO/IARC Trust Agreements, and may therefore be changed or even disregarded. In any case, they depend on the centres requiring and enforcing special agreements with recipients.

According to CIP guidelines, authorization for a recipient of designated germplasm to seek IPRs on a new variety can only be given by a centre if the new product (through an intellectual contribution) is significantly different from the originally transferred material (i.e. a product whose main characteristics and aptness are other than those of the initial material provided).[180]

Some breeders also are anxious for greater clarity in their dealings with germplasm provided by centres. They can accept wholeheartedly the obligation not to protect for their own benefit materials and information received from the centres. However, it is not clear to commercial breeders why they should be forbidden by CGIAR to apply for IPRs on materials that are significantly different. Guidelines such as CIP's are appropriate given the mandate of the centres to ensure that designated germplasm does not become encumbered by IP rights. Commercial breeders say that they will respect those restrictions. However, if they are unable to seek protection on any material derived from germplasm from gene banks, they will end up not using such germplasm. These policies, they claim, could therefore

See p. 111 *et seq.* for the Notes

jeopardize use and further development of designated germplasm. In the absence of IP protection, the investment will not be undertaken by industry because it could not be recovered — and valuable existing materials will be neglected. Potential benefits, they claim, are withheld from farmers and consumers for fear that breeders should make profits.

The FAO/CGIAR Trust Agreement specifically refers to germplasm and related information. The latter includes, where applicable, information on indigenous knowledge which should be protected according to Article 8j of the CBD. However, the MTA used by the CGIAR centres to transfer designated germplasm does not specifically refer to this obligation. The FAO/CGIAR Trust Agreement does not cover cells, organelles, genes, etc. from designated germplasm, although this is covered in the centres' IP guidelines and thus it is left up to the individual centre to enforce — without reference to the intergovernmental authority of the FAO.

Recent events demonstrate that it is difficult for the CGIAR centres to monitor recipients of designated germplasm. In addition, CGIAR centres lack mechanisms to counter violations of MTAs. Technically, gene markers on designated germplasm could be used to track utilization by third parties. Centres could, through FAO, alert governments in countries where violation has taken place. However, governments will not necessarily see it as their

Viewpoint box: Is the FAO-CGIAR Trust Agreement satisfactory?

Trust the breeding community

Despite decades of efforts to prove otherwise, the incidents of PBR or patent abuse or other forms of so-called 'biopiracy' represent an infinitesimal portion of the flow of germplasm and of breeding work. Instances where real commercial gain has resulted from dubious practices are still more unusual. Rather than devoting scarce resources to the pursuit of alleged 'pirates', we should be building trust within the international breeding community and encouraging the most unimpeded flow of germplasm possible. Any new arrangement must, first and foremost, strengthen the international public good associated with the fullest and freest possible gene flow.

Responsible Trust

After decades of unsung and costly work conserving agricultural biodiversity leading to the world's largest and most secure storage of unique crop germplasm, the CGIAR established a Trust Agreement with the intergovernmental community through FAO. Policy oversight now rests largely outside the CGIAR but the financial and material burden still remains with IARCs. It is time governments shouldered their full responsibilities and secured the future of these collections permanently.

Maintain the principles — strengthen the Trust

Two steps are needed: First, full legal ownership of the CGIAR collections should be transferred to the intergovernmental community through FAO. Second, a revised Trust Agreement (between FAO and governments) should define germplasm broadly, to include all elements of germplasm such as genes; and reaffirm that all of the germplasm encompassed by the trust accessions must remain in the public domain and cannot be privatized. The amended Trust Agreement should be incorporated into national laws, so that governments enforce it.

See p. 111 *et seq.* for the Notes

role to enforce the MTAs, which have only the status of private agreements. These problems deserve further consideration. Some would say it is not clear why it is vital for the centres to devote expensive resources to monitoring infringement of contractual IP rights arising from MTAs, yet inappropriate for them to take out other kinds of IP rights.

It remains unclear whether or not the parties to the Trust Agreement can construct an effective monitoring system that not only links information on trust germplasm to CGIAR use and exchange of that germplasm but also connects to national and international databases for PBR and patents.

Governments need to address the unresolved issue of *all* pre-CBD collections. The Nairobi Final Act, 1992, referred to the need to address all pre-CBD *ex situ* collections, meaning both domestic, national and international collections. Although the issue of rights over germplasm actually involves all pre-CDB *ex situ* collections held by any public or private institution in any foreign location, most of the international debate focuses on the CGIAR's gene bank material. Not only does the CGIAR have a vast and generally well-cared-for collection but it is the largest source of germplasm held by any 'international' body.

Some parties believe that any CGIAR material whose collection location is documented should be surrendered by IARCs to the country. The country may or may not wish to have the collection repatriated or duplicated. Among those who share this view are some who regard the FAO–CGIAR Trust Agreement as a political ploy engineered by the CGIAR even as the CBD was coming into force in order to avoid surrendering the germplasm to its rightful owners.

Some others insist that the actual history of the Trust Agreement makes it obvious that the CGIAR collectively and centres individually were forced by public pressure to accept the connection to FAO which they dearly wished to avoid. Indeed, it was the row that took place at the last meeting of states before the first COP (held in Nairobi in June of 1994) that literally railroaded the IARCs into accepting FAO authority over their collections. At that time, some governments in the CBD accused the World Bank of attempting to take over the IARC gene banks in order to bargain with the WTO for a special status for the CGIAR. The debate forced the CGIAR, they claim, to agree publicly to the Trust Agreement and to make the System rush to sign the agreement that October before the first COP session in Nassau in November.

Still others consider both of these scenarios to be too dramatic. The CGIAR's mission to increase food production, eradicate poverty and protect the environment in developing countries depends on the fullest possible exchange and use of plant genetic resources for food and agriculture (PGRFA). Furthermore, agrobiodiversity is a critical component of environmental conservation. The growing politicization and legal ambiguity posed a threat to the exchange and use of PGRFA and spurred the CGIAR

See p. 111 *et seq.* for the Notes

to look for a means to ensure that the collections be used for the benefit of the world community. Placing the collections under the auspices of a United Nations body — the FAO — and in trust for the world community was seen as the best mechanism available (pending finalization of the revision of the IU) to ensure the collections be deployed for the benefit of the world community. They believe the concept of repatriating gene banks is not a feasible option. It is technically complex because the germplasm held in genebanks may have developed its distinguishing characteristics somewhere other than the collection location. It is no simple matter to return germplasm to its 'rightful owner'.

There are at least three distinctly different policy positions today.

Viewpoint box: What about the CGIAR's pre-CBD material?

Waiting for a final solution
The CGIAR–FAO Agreement has to be understood as an interim step pending the revision and completion of the FAO's International Undertaking. The legal status of the CG's centres collection remains to be clarified and the centres should claim no property rights on the material in interim phase. At the present time, the centres also have no right to transfer their ownership to any entity. The collections must be managed as if they are part of a governance system which is still being defined by the International Undertaking and subject to the principles of the CBD as appropriate.

No change
Since the CBD has offered no solution to the question of germplasm collected and exported prior to its coming into force, the situation remains as it has always been. Unless there is a general agreement on universal repatriation of all germplasm from all sources (which seems impractical), it is difficult to understand why the world community would agree to repatriate germplasm that is being held for everyone's mutual benefit — especially given the difficulties in identifying the origin of much of the material.

Repatriate
The collection location of the majority of the germplasm in the CGIAR gene banks is known. Given the principle of sovereignty reaffirmed by the CBD and its recognition that pre-CBD collections are an unresolved issue, the resolution should logically be for IARCs to accede to the wishes of the governments involved and to repatriate the known material if requested to do so.

The long-term role of the CGIAR in access and exchange

Several events have come together to stimulate a very wide discussion on the future role that might be played by the CGIAR with respect to access and exchange. First, the Leipzig GPA clearly saw a central role for the CGIAR in delivering benefits to communities and countries through both germplasm conservation and enhancement work. Second, the CGIAR's own systemwide review (tabled by Maurice Strong in late 1998) prompted new thinking regarding the future of the CGIAR's genetic resources activities. Third, alongside the systemwide review, a special external review of the CGIAR's genetic resources program was undertaken. Where the reports

See p. 111 *et seq.* for the Notes

and interests will converge is not certain but many parties are beginning to express their views.

At the national level, some governments and scientists appear to see international centres as research-funding competitors. Some, too, look upon IARCs as a kind of 'fifth column' capable of receiving and shipping germplasm and undertaking initiatives that may not be compatible with national interests or policies. These views, on occasion, have led to calls to 'nationalize' or, at least, 'regionalize' the centres.

Others, in a world of diminishing financial flexibility, see the well-established centres as a — possibly 'the' — primary instrument for realizing the Leipzig GPA and bringing substance to the FAO multilateral exchange negotiations. Although many within the CGIAR would readily agree, some negotiators may be thinking in terms that would be far less attractive to the independent-minded IARCs. Should the intergovernmental community 'take over' the CG centres? Would such a move render the CG a hapless victim to bureaucracy and rent the finely woven fabric of scientific pursuits? Is there a middle ground? Does the newly created Global Forum on Agricultural Research offer an informal policy development environment that — together with FAO and the World Bank — offers *de facto* guidance to the System?

The role of the CGIAR will be debated intensely over the next few years. Whatever the outcome, there may be a growing consensus on a few key points. First, the current ambiguity with respect to CG-held germplasm collections must come to an end. Second, the legal status of each CG collection must be clear within the laws of the host countries and within the international community. Third, there is likely to be some subset of CGIAR germplasm which should be legally held by the international community.

Crucible Recommendation 11

Ensuring international gene bank security

The Crucible Group considers the network of international gene banks developed by the International Agricultural Research Centres and guided by the Systemwide Genetic Resources Program (SGRP) to be vital to future world food and agricultural development. The individuals and institutions that have created this living world treasure must be commended for their farsightedness and dedication.

Recognizing that the time has come to further strengthen this network and make the gene banks one possible element of benefit-sharing under the Leipzig GPA, the Group recommends that:

- the gene banks and the SGRP be brought under the aegis of the Conference of Parties of the revised International Undertaking so as to ensure the international legal and public character of the gene banks and a more stable and predictable flow of financial resources to them;
- the CGIAR centres and staff continue to manage and utilize the international gene banks as an integral part of their ongoing mandate activities.

See p. 111 *et seq.* for the Notes

Whether this subset is all of the present material held 'in trust' by FAO or is confined to germplasm whose point of collection cannot be identified remains to be seen.

CGIAR research and intellectual property

Even the CGIAR's toughest critics have conceded that the CG has moved both carefully and expeditiously to address the new environment created by the Biodiversity Convention and the renegotiation of the International Undertaking. The System's handling of suspected trust abuses has also been exemplary. Nevertheless, many continue to be distrustful of the CGIAR as centres struggle to sort out their differing approaches to intellectual property with respect to their own germplasm research. Can the CGIAR defend the integrity of trust germplasm from intellectual claims while pursuing their own claims with material derived from the trust collections? The debate on this issue arises later in the report.

Crucible Recommendation 12

Monitoring germplasm flows
The Crucible Group believes that maximum transparency in the maintenance and movement of genetic materials will create greater confidence in, and support for, a multilateral system. In order to achieve that transparency, the group recommends that:
- the germplasm and variety databases existing at international and national levels, such as the SINGER database, the UPOV database and the national catalogues, be made compatible;
- these databases be made openly accessible to all stakeholders.

See p. 111 *et seq.* for the Notes

Knowledge

To the surprise of many, national and international concerns related to the knowledge that resides with 'indigenous peoples' and farming communities — perennially emotive but invariably 'soft'-policy issues enjoying fickle public popularity — have demonstrated a staying power and political energy that could not have been anticipated even five years ago. The dimensions of any discussion involving indigenous peoples or the role of farming communities is guaranteed to be wide-ranging, complex, and highly charged. The Crucible Group acknowledges the importance and scope of these broader issues but has sought to address them within the framework of plant genetic resources. Its discussions have focused on the task of conserving and enhancing knowledge about biological diversity through both national policy and international treaties. The obvious central fora are the CBD, the FAO Commission (CGRFA) and the UN Human Rights Commission. Because much of the debate over 'knowledge' has been associated with intellectual property, the WTO and WIPO (World Intellectual Property Organization) are also discussed.

Regardless of the fora, there are important differences of opinion in several key areas that could impact on human rights, property, and trade. Some argue, for example, that indigenous peoples and farming communities are the creators and custodians of biological knowledge (particularly in medicine, agriculture and ecosystems). From this perspective, they argue, nation states and private enterprise should not file intellectual property claims that pre-empt or capture indigenous contributions. Some others, although appreciative of the primary and pivotal contribution of indigenous and other rural communities, believe that intellectual property systems properly acknowledge debts to past innovations and only claim new contributions. They also insist that sovereign states must be responsible for biological diversity. Others, still, identify a clear difference between the historic contributions of indigenous and farming communities and modern science. The debate has stimulated some governments, civil society organizations and intergovernmental agencies (such as WIPO) to review current intellectual property regimes to see if the knowledge of indigenous and farming peoples could be protected.

Much of the 'knowledge' debate connects with perspectives on human rights. Indigenous peoples often argue that their rights are being violated when their access to land and to benefits from their scientific contributions are ignored. Some have also insisted that Farmers' Rights (as discussed in FAO and the CBD) are Human Rights. This is especially contentious since FAO's negotiations appear to be moving to acknowledge Farmers' Rights

See p. 111 *et seq.* for the Notes

solely (or at least primarily) within national laws. Some believe that elements of Farmers' Rights involve the Right to Food and must be seen as standing above national laws. The slow pace of negotiations at the CBD and FAO with respect to these issues has led some observers to conclude that many developing countries are only using farmers and indigenous peoples as bargaining chips that can be quickly discarded if the North is prepared to offer sufficient financial or technological support.

There is no agreement — anywhere — on whether or not it is right or useful to incorporate indigenous knowledge into a formal intellectual property system. Advocates on all sides of this debate cite both practical concerns and issues of principle. Certainly, the Crucible Group was not able to reach agreement here.

Paradigm shift

Some advocates of 'informal innovation' believe that recent years have witnessed a paradigm shift in the recognition of and appreciation for the role of indigenous knowledge. There is new awareness among ('conventional', 'laboratory', 'Western', or 'institutional') scientists (no term is generally acceptable) that farmers and indigenous peoples not only have knowledge but often actively engage in research. Rather like the rediscovery of Mendel's Laws at the beginning of the 20th century, the end of the century saw a rediscovery of the creativity and innovation of rural societies. For some time, conventional science believed that indigenous knowledge was a hit-or-miss affair through which communities built up a storehouse of useful experiences passed from generation to generation. Conventional wisdom argued that this building-up of knowledge was erratic and imperfect. The scientific process championed by Isaac Newton and others was said to differ from traditional knowledge accumulation in its emphasis on experimentation and documentation.

Yet, scientific researchers (and Crucible members) such as Bo Bengtsson, Joachim Voss, Bernard Le Buanec and Louise Sperling — among others — have studied and assessed the contribution of farmers in Africa to the management of their genetic resources for food security and productivity. For example, they found that Ethiopian women would tabulate the yield results of their sorghum harvests on doorposts every year. They observed that women selected the highest-yielding, hardiest or otherwise most useful seeds from the field before men were allowed to harvest, and that these seeds would be 'tested' in kitchen plots and even exchanged with neighbors for trials in differing soils. This is experimentation and documentation. The energetic exchange of seed between farming communities, common around the world, was an effort to access diverse research material to improve food security. Whereas short decades ago, conventional science sometimes described farmers' varieties as 'Stone Age' or 'primitive', the last few years

See p. 111 *et seq.* for the Notes

have seen a shift in thinking and a much more realistic analysis of these varieties as ever-evolving and displaying differences from the seeds of the previous season.

As conventional science has adapted its language and approach to appreciate the experimental research of farming communities, it has also had to reconsider its understanding of what is 'known' and 'unknown', 'wild' and 'domesticated'. The discovery that a 'jungle' in West Africa was, in fact, an intentionally-developed agro-forestry system spurred the reappraisal of long-held assumptions. Many indigenous and farming communities eschew the term 'wild', for example, arguing that the term testifies to the limitations in the information available to conventional science. A so-called wild plant may be protected and nurtured if not actively bred. It is very often used and planted. Given the limitations of conventional understanding, they insist, the term 'wild' should only apply to species where scientists can prove there has been no prior protection or use. In the absence of this proof, the world should assume that it has been influenced by human intervention.

Some rural advocates are also sceptical of the quality of conventional claims of discovery and innovation. For many years it was thought, for example, that rubber had very limited use among indigenous peoples in the Americas because they had never developed a vulcanization process that would prevent it from becoming brittle. In fact, Charles Goodyear won a US patent in 1844 for his accidental discovery that a mixture of latex and sulphur wouldn't burn when placed on a stove and get improved elasticity.[181] Yet, recent investigations by scientists at Massachusetts Institute of Technology prove that Mayans had developed a latex vulcanization process of their own 3500 years before Goodyear's patent when they mixed the latex with juices from the morning glory vine.[182]

Knowledge accumulation today

So-called 'conventional' science has mixed views on this perspective. Many acknowledge the information and innovative capacities of rural communities with enthusiasm and are eager to engage in participatory plant breeding and other research initiatives with communities. (Some of these cases are identified in Part One of this volume. Readers should also note the related discussion in the second volume.)

Some institutional plant breeders point out that collaboration with farmers is really nothing new, it is precisely what institutional breeders have been doing for many years all over the world. After all, commercial breeders depend on farmers as their clients and customers.

Others regard the assertions of 'traditional' science as owing much more to political correctness and romantic idealism than reality. The rediscovery of Mendel's Laws, they suggest, led to advances in plant breeding that in 70 years have multiplied by several times the crop yields achieved over the previous nine millennia.[183] Others, while seeing the potential for collaboration between 'community innovation systems' and 'institutional innovation systems', describe the relationship differently. Communities, for some reason, often produce macro-system innovations that are most likely to be useful in specific micro-ecosystems (such as the farm or the community). These innovations often involve a complex of genetic and management improvements that rely, for their effectiveness, on the entire environment around them. Such improvements can be intentional and can have high value but they may not have wide application. Conversely, institutional innovators often create micro-innovations that have macro-applications — genetic manipulations that could find use in many ecosystems. Between these positions lie a multitude of variations.

Intellectual integrity

Some argue that present-day intellectual property claims on plant varieties or genetic traits and medicinal compounds are a usurpation of indigenous knowledge and an insult to the intellectual contribution of rural peoples. The extraction of a disease resistance gene from a farmers' variety that is known by the farmers to be disease-resistant, amounts to piracy even if the farmers had no scientific understanding of genes. Others strongly disagree, pointing out that the practical implications of such a position could be to curtail research and that the principal implication is that we should all still be paying royalties to the inventor of the wheel. Some farming and indigenous peoples' organizations counter by pointing out that they are not seeking recognition for past achievements but only for the evolutionary innovation in the field at the time that it was collected.

Despite the intensity of these differences, there is general agreement that traditional knowledge and knowledge systems need to be conserved and their further development encouraged. There is also general agreement that indigenous peoples and rural communities must take an active part, nationally and internationally, in policy formulation related to knowledge issues. Few, too, would deny that the creative collaboration possible among millions of researchers working in millions of test plots (i.e. farmers) in alliance with thousands of breeders in experimental stations, and many other kinds of researchers working in hundreds of laboratories (i.e. conventional scientists), would not be a boon to knowledge and diversity.

The Crucible Group examines three outstanding issues related to knowledge in the following pages:
1. Human rights and indigenous knowledge;
2. Participation of indigenous and local communities in knowledge policy-making;
3. Intellectual property and indigenous/local communities.

See p. 111 *et seq.* for the Notes

Human rights and indigenous knowledge: potential role for the UN Human Rights Commission

Policy primer

- With headquarters in Geneva, the UN Human Rights Commission is the United Nations' main policy-making body dealing with human rights issues. Composed of 51 member countries in its governing body, it prepares studies, makes recommendations and drafts international human rights conventions and declarations.
- Over the past decade, the Commission's Working Group on Indigenous Populations has played a lead role in the articulation of indigenous peoples' rights. The Draft UN Declaration on the Rights of Indigenous Peoples gives important recognition to the rights of indigenous peoples over cultural and genetic resources.
- Following a decision at the World Food Summit in 1996, the Human Rights Commission has been working with FAO and other parties to review the various international agreements related to 'The Right to Food'.
- With the appointment of Mary Robinson (former President of Ireland) as the new United Nations High Commissioner for Human Rights in June 1997, the Commission has attained a high level of credibility and influence.

The Universal Declaration of Human Rights adopted by the UN General Assembly in 1948 recognizes both the right to innovate and the right to imitate. While the Declaration is almost universally admired as a landmark contribution to democracy and development, it has nevertheless been criticized as a somewhat Western-centric perspective that stresses individual rights and freedoms and does not adequately address cultural and collective rights, especially those of indigenous communities. The long succession of protocols and deliberations since 1948 could be seen, to some degree, to acknowledge this limitation. The Crucible Group reviewed two issues that involve the work of the UN Human Rights Commission.

Indigenous rights

Although there is no agreement within the Group on this point, some members consider the issue of indigenous knowledge protection to be a human rights matter that should be resolved by the UN Human Rights Commission — not by WIPO, WTO or the CBD. Those holding this opinion suggest that other parties and treaties should defer their conclusions until the Commission can provide guidance.

Farmers' Rights/the Right to Food

Second, as already noted under the Farmers' Rights discussion earlier in Part Two, some negotiators believe that there are elements of the debate in the FAO International Undertaking with respect to Farmers' Rights that could be more substantially addressed within the framework of the Right to Food if considered by the Human Rights Commission. A June 1999 study by the Economic and Social Council (ECOSOC) on the Right to Food submitted

See p. 111 *et seq.* for the Notes

to the Commission on Human Rights urges that Farmers' Rights should be promoted as part of the Right to Food, especially since 'our future food supply and its sustainability may depend on such rights being established on a firm footing'.[184]

The participation of indigenous and local societies in knowledge policy-making

The importance of ensuring the participation of indigenous and local people in negotiations that will lead to national or international policies affecting their knowledge is obvious. The difficulty arises in the relations between indigenous peoples and national governments. Some indigenous peoples understand themselves to be a nation within a nation or a nation whose peoples cross the borders of two or more nations. Some governments consider themselves to be the sole and entirely sufficient voice of all the peoples within their sovereign territory. In a world of inherited colonial (or victor-dictated) boundaries and disinherited societies, it is easy to sympathize with all points of view. For this reason, many national and international organizations have adopted flexible and often ambiguous procedures intended to encourage inclusive policy formulation. Nowhere is this flexibility more appropriate than when addressing the knowledge related to biological resources that have themselves experienced a kind of diaspora.

For guidance on this issue, it is important to consider the Draft UN Declaration on the Rights of Indigenous Peoples. This is a comprehensive articulation of indigenous peoples' rights, which was developed over the past decade by the Working Group on Indigenous Populations. In addition to rights to political and legal autonomy, the Draft Declaration gives important recognition to the rights of indigenous peoples over cultural and genetic resources. Some governments, like that of the Philippines, with its recently passed Indigenous Peoples' Rights Act, have attempted to create national laws that give effect to the spirit of Article 29 of the Draft Declaration.

Although the declaration is still in draft form, it includes provisions that are internationally recognized as minimum standards in the field of indigenous rights. Article 29 states:

> Indigenous peoples are entitled to the recognition of full ownership, control and protection of their cultural and intellectual rights.... They have the right to special measures to control, develop and protect their sciences, technologies and cultural manifestations, including human and other genetic resources, seeds, medicines, knowledge of the properties of fauna and flora, oral traditions, literatures, designs and visual and performing arts.

If and when it is adopted, the Draft Declaration on the Rights of Indigenous Peoples will not be legally binding but will give the world a powerful moral lead.

International attention has, logically, focused on the involvement of

See p. 111 *et seq.* for the Notes

indigenous peoples in the Biodiversity Convention. It could also be argued that their participation is as important wherever there are proposals to commercialize knowledge — including the World Intellectual Property Organization and the WTO.

Crucible Recommendation 13

The rights of indigenous peoples with respect to genetic resources
The Crucible Group recommends that national governments should adopt, within their own laws, the spirit and the literal text of Article 29 of the Draft Declaration on the Rights of Indigenous Peoples.
As means of further implementing the spirit of Article 29 into national law, the Crucible Group recommends that national governments should:
- give legal recognition to the collective rights of indigenous peoples and local communities and to their knowledge in relation to genetic resources;
- recognize customary and traditional practices that indigenous peoples and local communities have developed regarding genetic resources; and
- require by law that third parties should not be able to assert rights or make claims with respect to plant varieties and uses of medicinal plants developed by indigenous and local communities without their prior informed consent.

Participation in the CBD
The CBD gives international recognition to the central role that indigenous and local communities play in biodiversity conservation, through their traditional and sustainable practices and knowledge systems. The CBD encourages signatory states to protect and promote the rights of communities, farmers and indigenous peoples vis-à-vis their biological resources and knowledge systems. It also encourages the equitable sharing of benefits arising from the commercial use of communities' biological resources and local knowledge:

> Each Contracting Party shall, as far as possible and as appropriate [...] subject to its national legislation, respect, preserve and maintain knowledge, innovations and practices of indigenous and local communities embodying traditional lifestyles relevant for the conservation and sustainable use of biological diversity and promote their wider application with the approval and involvement of the holders of such knowledge, innovations and practices and encourage the equitable sharing of the benefits arising from the utilization of such knowledge, innovations and practices.
>
> *The Convention on Biological Diversity, Article 8(j)*

The CBD has become a strong and significant forum for indigenous peoples and local communities, whose representatives have participated actively in negotiations and debate. Indigenous peoples have lobbied

See p. 111 *et seq.* for the Notes

forcefully at meetings of the Conference of the Parties (COP) to ensure that they would have a more prominent place within the CBD. In November 1997 the COP held an open-ended inter-sessional working group discussion on traditional knowledge and biological diversity in Madrid. The workshop produced recommendations concerning, *inter alia*, a program of work related to the implementation of Article 8(j). In May 1998, COP IV established an open-ended working group to address the implementation of Article 8(j) and related provisions.

There are different viewpoints on the nature of indigenous peoples' participation at the CBD, and whether or not it is a useful forum for advancing their rights. Some indigenous peoples' organizations question whether multilateral agreements negotiated among nation states, emphasizing national sovereignty over genetic resources, will adequately consider or protect the rights of indigenous peoples. Many view the adoption of the Draft Declaration on the Rights of Indigenous Peoples by the UN General Assembly as a far more important priority in the intergovernmental arena. Some view the CBD as a treaty that promotes the commercial exploitation of biological resources under the guise of an environmental treaty, and are skeptical about its role in protecting the rights of indigenous peoples.

Viewpoint box: Is the CBD a useful forum for advancing the rights of indigenous peoples?

One step forward
The CBD is a legally binding document, unlike the Draft UN Declaration on the Rights of Indigenous Peoples, which is a 'soft' law. Article 8(j) is a victory for indigenous peoples because it acknowledges the need to promote and protect the knowledge of indigenous peoples in biodiversity conservation and sustainable use. Since it is legally binding, contracting parties are required to comply with its decisions. Indigenous peoples have to monitor closely how their governments are complying.

While the implementation of Article 8(j) will not comprehensively address indigenous peoples' issues, it is already a major building block in the struggle for indigenous peoples' rights and self-determination.

Two steps back
CBD is a distraction from the bigger issues of territoriality and human rights. The CBD is not of much use to indigenous peoples because it is based on the principle of national sovereignty. And it emphasizes the commodification of biological resources under the guise of an environmental treaty. The de-contextualization of biodiversity from culture and rights to territory must be challenged. The Draft UN Declaration on the Rights of Indigenous Peoples enjoys broad support from indigenous peoples and it is the most comprehensive articulation of indigenous peoples' rights — including the right to have control over territories and resources. This is the framework upon which all other rights should be based. Indigenous peoples should not be distracted from their focus of ensuring that the Draft Declaration gets adopted by the UN.

See p. 111 *et seq.* for the Notes

Viewpoint box: What should be indigenous peoples' representation at the CBD?

Driver's seat
Indigenous peoples should be recognized as full parties to all international fora addressing their concerns (especially but not exclusively the CBD), because only then will their interests be protected and promoted. Governments cannot adequately represent indigenous peoples' interests because, in most cases, they are the key violators of indigenous peoples' rights. The main reason why indigenous peoples bring their concerns into the international arena is that they are ignored and marginalized at the national level. The imperative for governments and the international community is to recognize and promote the rights of indigenous peoples.

Back seat
Indigenous peoples can be represented in various ways:
1) Governments shall include indigenous persons in their delegation;
2) They shall attend as observers with participation in the widest possible extent in the deliberations;
3) Secretariats can choose indigenous peoples who will be considered as experts and they could become members of a subsidiary body. However, they cannot be parties to treaties and will be regarded as observers just like all non-state actors. Indigenous peoples can widen the space given to them once they are inside. Thus, even if this is a back seat, it still provides an opportunity where indigenous rights can be advanced.

Intellectual property and indigenous/local communities

Thorniest of all issues in the knowledge debate is whether or not legal mechanisms can or should be created to ensure protection of the knowledge of indigenous and local people. Some believe that this knowledge should be protected under intellectual property regimes modified, if necessary, for the purpose. Others assert that the inclusion of indigenous innovations in 'Western-style' IP regimes would not only not create new benefits but would actually camouflage biopiracy and amount to a slippery slope toward greater corporate monopolies and the end of opposition to the patenting of life. Some who share this opinion are campaigning for an alternative rights regime that is distinctly *not* intellectual property. They reason that a system could be devised that is compatible with traditional practices in each community and allows the community to govern access to its knowledge. Still others believe that alternative rights regimes are just a more naïve form of slippery slope leading inevitably to patent models and transactions. They suggest that efforts to legislate indigenous knowledge run against customary practices and threaten the survival of cooperative innovation systems. The World Intellectual Property Organization is in the middle of this debate.

For a more detailed discussion of this issue, readers should refer to the viewpoint box on suitability of IP for indigenous knowledge in Volume 2.

See p. 111 *et seq.* for the Notes

The potential role of WIPO in protecting indigenous knowledge

Policy primer

- Established in 1974 with headquarters in Geneva, WIPO has 171 member states and is responsible for the promotion and protection of intellectual property throughout the world. WIPO administers 19 international treaties concerning intellectual property.
- WIPO cooperates with the WTO and offers legal and technical assistance to developing and other countries concerning the implementation of the TRIPs Agreement.
- In 1997 WIPO created the Global Intellectual Property Issues Division, which is exploring the intellectual property needs of holders of traditional knowledge, innovations, culture and genetic resources.

Outstanding issues:
- Can and should the knowledge of indigenous and local communities be afforded formal intellectual property protection?
- If so, how should this knowledge be recognized nationally and internationally and how will its recognition impact upon other intergovernmental agreements?

A new initiative at the World Intellectual Property Organization (WIPO) acknowledges that the needs of informal innovators have not been addressed by current IP regimes and seeks to address this situation. Over the past five years, representatives of indigenous peoples' organizations have participated actively in debates relating to IP and genetic resources at regional and international fora. In 1997 WIPO created the 'Global Intellectual Property Issues Division (GIPID)'. For the 1998–99 biennium, the GIPID focused on: 1) the intellectual property needs of new beneficiaries; 2) biodiversity and biotechnology; 3) expressions and folklore; and 4) IP beyond territoriality. WIPO is in the process of exploring both current approaches to the protection of IP and indigenous knowledge, as well as options for new or adapted forms of protection.

In 1998–99 WIPO conducted a series of regional fact-finding missions to North America, South and Central America, the South Pacific, West and Southern Africa and South Asia, with the aim of identifying and exploring 'the intellectual property needs and expectations of new beneficiaries, including the holders of indigenous knowledge and innovations, in order to promote the contribution of the intellectual property system to their social, cultural and economic development.' WIPO will release a full report on its fact-finding mission in early 2000. WIPO also invited indigenous peoples to participate at two Roundtables on Intellectual Property and Traditional Knowledge. The first was held in July 1998, the second in November 1999.

Some indigenous peoples' organizations question WIPO's objectives and the assumption that holders of traditional knowledge have something to gain from existing IP regimes. Some indigenous leaders fear that predominantly Western-style IP regimes could undermine local knowledge systems, rather than promote or protect them. Others believe that there might be a role for WIPO to play in helping to establish alternative forms

See p. 111 *et seq.* for the Notes

of IP. There is considerable difference of opinion on whether or not IP regimes can or should be used to protect traditional knowledge, but there is general agreement that the issue of traditional knowledge should be addressed by WIPO and other elements of the international community as a step toward implementing Article 8(j) of the CBD.

Other IP-based efforts outside WIPO

Many national governments and peoples' organizations are working to develop IP-based protections for indigenous and local knowledge. Many of these efforts predate the WIPO initiative.

Some CSOs and policy-makers are working to develop community and collective rights regimes for farmers and indigenous peoples that are distinct and separate from monopoly-based, intellectual property regimes. For example, CSOs from 19 countries met in Thailand in late 1997 to explore the development of *sui generis* rights that aim to recognize and to protect community innovation, and also nurture sustainable food and health systems.[185] 'The Thammasat Resolution' that emerged from that meeting asserts that *sui generis* rights are community and collective rights that are fundamentally different from the intellectual property-based *sui generis* systems promoted by the TRIPs Agreement. Participants rejected IPRs on all forms of life.

One of the earliest comprehensive efforts to develop model *sui generis* national legislation that would give communities property-style rights of control over their collective knowledge was the Third World Network's Community Intellectual Rights Act (CIRA), published in 1994. The Act addresses many of the issues that continue to challenge current efforts to develop laws to protect indigenous and local knowledge. It asks, for example: what subsets of knowledge can be protected by such laws? In whom should the rights of protection be vested? Can more than one community have rights in the same knowledge? What kinds of rights of control should be created in association with that knowledge? How long should those rights last? Certainly not everyone agrees with the way in which the CIRA addresses all of these issues. The Act is not long or detailed enough to deal comprehensively with all of the questions it raises. However, it represents an early and influential attempt to begin framing, in legal and legislative terms, ways in which indigenous and local knowledge could be treated in national *sui generis* intellectual property laws. (Please refer to Volume 2, Topic two, Section 2 — 'Options for *sui generis* intellectual property laws for indigenous and local knowledge' — for more detailed discussion of *sui generis* laws that seek to protect local and indigenous knowledge.)

At WTO, a number of developing countries are raising the issue of protection of indigenous and local peoples' rights over their collective knowledge. Kenya, on behalf of the African Group at WTO, submitted proposals relating to the TRIPs Agreement in preparation for the 1999

Ministerial Conference.[186] Among them was a proposal that any *sui generis* law for plant variety protection may provide for the protection of the innovations of indigenous and local farming communities in developing countries, consistent with the Convention on Biological Diversity and the International Undertaking on Plant Genetic Resources. A joint proposal by the governments of Peru, Bolivia, Colombia, Ecuador and Nicaragua in October 1999 calls upon the WTO to study and make recommendations on the most appropriate means of recognizing and protecting traditional knowledge as the subject matter of IP.[187] It also calls upon the WTO to establish in upcoming negotiations a multilateral legal framework 'that will grant effective protection to the expressions and manifestations of traditional knowledge'.

Much of the work regarding indigenous and local knowledge overlaps with ongoing work on *sui generis* forms of IP protection for plant varieties. One possibility discussed in the context of *sui generis* plant variety protection regarding indigenous and local knowledge is that some form of IPRs could be extended to plant varieties developed and grown by indigenous and local farmers. For further discussion, refer to the section entitled 'Options for the implementation of Article 27.3(b)'.

What unites most of the efforts of groups outside of the WIPO initiative (it is still too early to know what WIPO will recommend or observe) is the desire simultaneously to recognise the collective aspect of indigenous and local community stewardship and to provide those communities with different forms of control (ranging from restrictive to exclusive) over their knowledge.

There is disagreement over the characterization of some of these efforts. Some critics talk about models of protection that vest exclusive or restrictive rights of control over knowledge in a community as being systems that are outside of, or distinct from, intellectual property law. Others claim that any legally sanctioned rights of control that can be asserted against third parties, independent of whether they are vested in individuals or communities, are forms of intellectual property (albeit *sui generis* ones), and that political correctness is preventing some critics from acknowledging this.

If formal intellectual property procedures were modified, would it adequately safeguard the interests of indigenous and other local communities? If indigenous knowledge were properly considered at intellectual property offices, some people believe that disputed claims related to medicinal plants and plant varieties (i.e. ayahuasca, neem, basmati) might be much reduced or even disappear.

In November 1999 the US Patent and Trademark Office (PTO) cancelled a US patent issued to a US citizen for a plant species native to the Amazon rainforest, *Banisteriopsis caapi*. Popularly known as the ayahuasca vine, the plant is used in sacred indigenous ceremonies throughout the Amazon. The PTO's decision came in response to a request for re-examination of the

See p. 111 *et seq.* for the Notes

Viewpoint box: Is intellectual property protection for indigenous knowledge and rights appropriate?

Not appropriate
The notion of retroactive protection for old ideas that may, or may not, have originated from one or more communities (or even countries) is absurd. IPRs has always offered a limited monopoly for new ideas. In general, ideas that have already been commercialized, or even published, cannot be protected. If protected, the protection lasts only for about two decades — not two millennia. We'll have the discoverers of indigo dye suing Leonardo da Vinci for the Mona Lisa while the descendants of da Vinci sue Madonna for trademark infringement!

Legal recognition of knowledge is always appropriate
Ideas and/or the expression thereof should be protected, irrespective of their being originated by indigenous and/or local people or multinational corporations. The legal regimes may be differentiated, in order to be adapted to their respective subject matter, but the economic and ethical principles are the same.

Evolutionary innovation
No one is trying to patent fire. Modern IPRs are merely acknowledging that different forms of innovation can exist and that, with respect to biological materials, individuals and communities are continuing to innovate medicines, preparations and agricultural varieties.

'Evolutionary innovation' implies that each generation of the bio-material has been improved over the previous generation in the same way that a commercial breeder establishes a key cross and then develops successive varieties from the cross over many years. The levels of protection for these different levels of innovation can also vary significantly.

Intellectual integrity — no exclusive monopoly
The task is not to protect specific communities but to safeguard the free flow of ideas and biological resources within global society. Therefore, intergovernmental bodies must ensure the intellectual integrity of those who claim innovations. Such persons must be able to 'prove' beyond reasonable doubt that they have actually contributed something new and useful to humanity. Under no circumstances should inventors be allowed exclusive monopoly over their inventions for any period of time. Life forms should not be subject to exclusive monopoly claims. After all, these are not inventions but discoveries.

patent in March 1999 by the Washington DC-based Centre for International Environmental Law (CIEL), the Coordinating Body of Indigenous Organizations of the Amazon Basin (COICA) and the Amazon Coalition. The groups requested that the patent be cancelled 'because the claimed patent lacks novelty and distinctiveness, is found in an uncultivated state, and as a sacred element of many indigenous cultures of the Amazon should not be subject to private appropriation'.[188]

In response to the PTO's cancellation of the ayahuasca patent, CIEL attorney David Downes observed that the PTO had still not dealt with the flaws in its policies that made it possible for someone to patent this plant in the first place. He called for the PTO to change its rules to prevent future patent claims based on the traditional knowledge and use of a plant by indigenous peoples.[189]

In separate proceedings at the PTO, the three groups have called for changes that would require that PTO patent applicants identify all biological resources and traditional knowledge that they used in developing the claimed invention, disclose the geographical origin of the claimed biological resources and provide evidence that the source country and indigenous community consented to its use.

Crucible recommendation 14

An intellectual property 'ombudsman'
The Crucible Group recommends, as a partial contribution to issues relating to indigenous and local people, that:
- WIPO and UPOV establish an ombudsman's office, accessible to indigenous and rural communities to address their queries and concerns regarding issues that arise within the competence of those organizations. The ombudsman should be empowered to achieve resolution of any issues that the office regards as pertinent to pursue;
- in further recognition of the importance attached to the participation of indigenous and local communities by the CBD and the support expressed for Farmers' Rights in FAO, a permanent ombudsman's office be created in the United Nations (possibly as part of a permanent Forum for Indigenous Peoples) to address a broader range of knowledge-related concerns than can properly be addressed under WIPO or UPOV;
- such offices should receive adequate financial resources and technical support to ensure that they are effective and functional.

See p. 111 *et seq.* for the Notes

Innovation

Governments around the world share a common concern for the need to stimulate scientific and technological innovation. As Volume Two of this report points out, national policy-makers can encourage innovation through a number of policy and legislative mechanisms ranging from general support for higher education and collaborative research to more targeted instruments enabling community science or intellectual property. As already identified in earlier sections, however, much of the policy friction centres on the international intellectual property conventions associated with WIPO and the Union for the Protection of New Varieties of Plants (UPOV). Also, as already noted, the role of the WTO and its TRIPs chapter is seen as, at once, crucial and controversial.

There is a debate (considered irrelevant by some and important by others) surrounding the impact of intellectual property on national development. Some view the history of IP as an attempt to deny innovations to 'have-not' countries. They argue that a country's view of IP changes when it shifts from the 'have-not' to the 'have' category. Others insist that intellectual property has been the leading engine in stimulating the unprecedented scientific progress of this century.

Within the IP debate on innovation are other concerns. Should farmers have the right to save and exchange proprietary seeds? Should scientific researchers have access to patented products or processes for the purpose of developing new plant varieties? Are some patent claims too broad? Should the current WTO TRIPs flexibility with respect to *sui generis* systems be maintained? Each of these issues can be argued from at least three different sides and each raises fundamental questions about the future wellbeing of the world's genetic resources.

In this section, the Crucible Group examines the following outstanding issues:

1. The Agreement on Trade-Related Aspects of Intellectual Property Rights;
2. The implementation of Article 27.3(b);
3. Alternative forms of plant variety protection that may implement Article 27.3(b);
4. The right of farmers to save and exchange proprietary seeds;
5. Review of Article 27.3(b);
6. CGIAR and intellectual property.

See p. 111 *et seq.* for the Notes

At the heart of many issues — the World Trade Organization (WTO)

Policy primer

- The WTO is based in Geneva and currently has 134 signatory states to its 1995 agreement. It is the world's most powerful intergovernmental trade regulator.
- The WTO operates on the basis of one country/one vote.
- Attempts to launch the 'Millennium' trade round failed at the Seattle ministerial meeting in November/December 1999. It is too early to say when or if member states will try again to get the negotiations started.
- The WTO today, through TRIPs, administers the most comprehensive multilateral agreement on intellectual property.
- Under the TRIPs Agreement, all members are obligated to adopt minimum standards for intellectual property rights and a mechanism for their enforcement.
- The TRIPs Agreement requires member countries to make patents available for inventions, whether products or processes, in all fields of technology. However, plants and animals may be excluded from patentability. For plant varieties effective protection shall be provided by patents and/or by an effective *sui generis* system.

Outstanding issues:

- Does TRIPs negatively or positively affect the conservation and utilization of biological diversity and do signatory states face a conflict between their efforts to comply with TRIPs and the CBD?
- Should TRIPs Article 27.3(b) be maintained or amended? Should exclusions from patentability be eliminated, or should exclusions be expanded to give members the option of excluding all biological materials from patentability?

The Agreement on Trade-Related Aspects of Intellectual Property Rights (TRIPs)

Historically, intellectual property laws are rooted in the principle of national sovereignty. IP was a domestic policy issue that was based on each country's level of development and technological needs. Until recently, the World Intellectual Property Organization (WIPO) was considered the most important intergovernmental forum on intellectual property rights. All of that changed with the creation of the World Trade Organization (WTO) in 1994. One reason is that WIPO's treaties on IP include few mechanisms for enforcement, dispute settlement, or compliance. By contrast, the WTO places far more pressure on countries to adopt minimum standards of IP; member countries must assume the obligations of WTO Agreements (including intellectual property) in order to gain WTO membership. The WTO's dispute settlement procedure creates a strong mechanism for compliance, including the power to impose trade sanctions against member states that fail to abide by its binding agreements.

The World Trade Organization today administers the most comprehensive multilateral agreement on IP. WIPO continues to provide technical assistance on national IP laws and institutions and administers 19 international treaties concerning intellectual property (such as the Paris Convention on Protection of Industrial Property and the Berne Convention on Protection of Literary

See p. 111 *et seq.* for the Notes

and Artistic Works). But although it remains an important body for international standard setting on IP, it is overshadowed by the more powerful WTO.

1994 saw the conclusion of the Uruguay Round of the General Agreement on Tariffs and Trade (GATT) and the creation of the World Trade Organization (WTO) which came into being in January 1995. The WTO operates on the principle that a liberalized system of international trade based on non-discrimination and the elimination of trade barriers is essential to global economic well-being. The WTO is based in Geneva; as of May 1999, 134 countries were members. The WTO's primary functions are to:

- administer WTO trade agreements;[190]
- act as a forum for multilateral trade negotiations;
- handle trade disputes;
- monitor national trade policies;
- provide technical assistance and training for developing countries; and
- cooperate with other international organizations.

What is TRIPs?

The Agreement on Trade-Related Aspects of Intellectual Property Rights (TRIPs) is annexed to the WTO Agreement. TRIPs brings intellectual property to centre stage in multilateral trade negotiations. Prior to the Uruguay Round, the GATT dealt mainly with trade in merchandise, and was not concerned with services and intellectual property. The inclusion of IP reflects, in part, the explosive growth in information technology and biotechnology in international trade, and the strong desire of some industrialized countries, particularly the US, Europe and Japan, to protect products from intellectual 'piracy' in foreign markets. Between 1980 and 1994 the share of high technology products in international trade doubled, from 12% to 24%.[191]

The TRIPs Agreement ensures that all signatories provide minimum standards of protection in a number of different areas of intellectual property law (e.g. patent, copyright, geographical indications, and so on). In this way, TRIPs makes these minimum standards universal, at least as far as the cumulative body of signatory states is concerned. The TRIPs Agreement requires member countries to make patents available for inventions, whether products or processes, in all fields of technology without discrimination, subject to the standard patent criteria of novelty, inventiveness and industrial applicability. TRIPs requires that patents be available without discrimination as to the place of invention and whether products are imported or locally produced (Article 27.1). However, TRIPs does not aim at identical national IP laws, nor does it establish an international patent system; rather, it sets minimum standards for member countries to follow. Under the WTO TRIPs Agreement, all members are obligated to adopt minimum standards for intellectual property rights and a mechanism for their enforcement.

Of particular relevance to biological diversity, TRIPs Article 27 requires that all member states adopt national-level IP systems for all products and processes, including pharmaceuticals, modified microorganisms and microbiological processes. During the negotiations, however, members had a difficult time reaching consensus on the controversial area of biotechnological inventions. While some industrialized countries pushed for no exclusions to patentability, some developing country members preferred to exclude all biological diversity-related inventions from IP laws. Still other members preferred something in the middle.

The text that prevailed is found in Article 27.3(b). It states that plants and animals as well as essentially biological processes may be excluded from patentability. However, WTO members must offer protection for plant varieties either by patents and/or by an effective *sui generis* system. *Sui generis* means a system of rights that is unique, 'of its own kind', for a specific item or technology.

However, the Agreement on TRIPs does not define *sui generis*. Although several attempts have been made, the term has not been defined in TRIPs negotiations.

Because of the difficulty in reaching consensus, it was agreed that the controversial sub-paragraph would be reviewed in 1999. Of course, 1999 has come and gone, and the review of 27.3(b) did not get started. It is hard to predict at this point when it will get addressed, and in what form. No doubt a great deal will be said about the effect of article 27.3(b)'s current wording in the context of the general review of the implementation of TRIPs scheduled to take place in 2000 (as mandated by article 71.1). In the end, however, WTO member states may find themselves too tied up with other business to be able to directly address the possibility of amending 27.3(b) in the next few years.

TRIPs Article 27.3(b)

Members may also exclude from patentability:

(b) plants and animals other than micro-organisms, and essentially biological processes for the production of plants and animals others than non-biological and micro-biological processes. However, Members shall provide for the protection of plant varieties either by patents or by an effective *sui generis* system or by any combination thereof. The provisions of this sub-paragraph shall be reviewed four years after the entry into force of the WTO Agreement.

Developing countries have until 2000 to pass laws in this direction, and least-developed countries (LDCs) have until 2006.

Ordre public *or morality*

Article 27.2 allows the exclusion from patenting of inventions contrary to *ordre public* or morality; this explicitly includes inventions dangerous to human, animal or plant life or health or seriously prejudicial to the environment.

See p. 111 *et seq.* for the Notes

Viewpoint box: Can TRIPs' public morality exclusion be used to reject patents on life forms or controversial new technologies such as genetic seed sterilization?

Inventions the use of which is necessary to be banned may be excluded
TRIPs does not require an actual ban of the commercialization of an invention as a condition for its exclusion from patent protection; such a ban only has to be 'necessary' to protect *ordre public*/morality. Otherwise, many inventions, especially those nobody had thought of before, and which therefore are unregulated would pass the morality test, whether their exploitation ought to be prevented or not.

It's a non-issue
Patents give a right to exclude others from using the patented invention without the patent holder's consent; they give no positive right actually to use an invention. Patent law and lawyers should therefore not have to deal with the morality of an invention. Whether an invention may be used or not should be left to other fields of law, such as biosafety legislation.

No rewards for immoral inventions
It is said that patents are intended to act as incentives for intellectual efforts, for financial investments and as reward for the disclosure of knowledge that would otherwise be kept secret. Patent law should not reward immoral inventions. It should not reward the disclosure of knowledge about such inventions. And states should not set incentives for the development of such inventions.

Inventions the use of which is not banned must not be excluded
The TRIPs Agreement prohibits WTO members from refusing a patent for an invention whose application, or at least commercial exploitation, is not even prohibited by their national laws, such as biosafety laws. Even if prohibited under these laws, the exclusion must not be made merely because the exploitation is prohibited.

This exclusion is subject to the condition that the commercial exploitation of the invention must also be prevented and that this prevention must be necessary for the protection of *ordre public* or morality.

There is disagreement among Crucible members as to the scope of this optional exclusion and its application to specific technological developments, such as genetic use restriction technology (GURT).

At a time when the global economic system is facing severe crisis, the short-term cost of implementing TRIPs, including considerations such as structural capacities, financial resources and technical expertise, is a significant factor for many LDCs. A 1996 study by UNCTAD estimates that, in order to comply with the TRIPs Agreement, Bangladesh will need to spend $250 000 in one-time costs for legislative drafting and over $1.1 million in annual costs for judicial work, equipment and enforcement measures — not including the cost of training personnel.[192] For the United Republic of Tanzania the costs of implementing TRIPs are estimated at $1–1.5 million.

There is concern that implementation of TRIPs could divert scarce resources away from basic social programs in the developing world. Article

67 of TRIPs calls for bilateral technical assistance to be given by industrialized countries to developing countries, which might help to reduce some of these costs. African countries, in particular, have expressed concern that not enough has been done under Article 67 to provide technical assistance for the implementation of the TRIPs Agreement.[193]

The implementation of Article 27.3(b)

Options for implementing Article 27.3(b)
The obligations imposed by the TRIPs Agreement require many developing countries to enact IP legislation for plant varieties (and other biological materials) for the first time. With deadlines for implementation and enforcement of new legal regimes fast approaching, members have a finite time period to explore options and implement patent and/or *sui generis* systems for plant variety protection. The TRIPs Agreement does not specifically define what an 'effective *sui generis* system' might look like, and thus allows for many different systems to emerge.

Discussion on how to implement Article 27.3(b) has tended to focus on the following questions: Is it best for developing countries, most of which have not yet offered any IP protection for plant varieties, to follow the model of plant breeders' protection as foreseen in the UPOV Acts? Or should developing countries explore alternative protection systems for plant varieties rather than following the ready-made protection systems currently being used in many industrialized countries? Which of the different options will better serve a country's particular national interests? It is in the context of this question that we examine UPOV and the UPOV Conventions, and then other, alternative forms of plant variety protection.

Although some argue that the existing system of plant breeders' rights is by far the best way to implement the TRIPs obligation to provide for the effective protection of plant varieties, it is recognized that the TRIPs Agreement does not oblige WTO member states to adopt a protection system along the lines of existing Plant Breeders' Rights. There is also agreement that WTO member states are under no obligation to become members of UPOV.

The Union for the Protection of New Varieties of Plants (UPOV)
Founded in 1961, UPOV is an intergovernmental body that establishes international rules under which countries grant intellectual property rights to the developers of new plant varieties (individuals or institutions). To qualify for protection, a new variety must be novel, distinct, uniform and stable. The original UPOV Convention was revised in 1972, 1978 and 1991. Today, the vast majority of UPOV members are party either to the 1978 or the 1991 Act.[194]

Historically, UPOV membership consisted primarily of a handful of industrialized nations. In recent years, that has begun to change. With the

See p. 111 *et seq.* for the Notes

Policy primer

- UPOV is an intergovernmental body that establishes international rules under which countries grant intellectual property rights to the developers of new plant varieties.
- UPOV is gaining prominence as a legislative model for Plant Breeders' Rights because Article 27.3(b) of TRIPs obligates WTO members to adopt patents and/or 'an effective *sui generis* system' for plant varieties.
- In April 1998, the 1991 Act of the Union for the Protection of New Varieties of Plants (UPOV) Convention entered into force. This was supposed to close the door on the 22 year-old 1978 accord, making it impossible to adhere to the 1978 Act from that date onwards. The UPOV Council, however, has made a derogation for those countries which sent UPOV their draft laws for review before the closing date. As a result, India, for example, has the opportunity to adhere to UPOV 1978 instead of UPOV 1991.
- The 1991 Act does not mandate an exemption allowing farmers to freely use farm-saved seed as further planting material. It does, however, leave each state free to include the farmer's exemption (or farmer's privilege) in national legislation.

Outstanding issue:
- Should farming communities have the right to replant or exchange farm-saved seed protected by intellectual property?

recent accession of the People's Republic of China, Kenya, Bolivia, Brazil and Slovenia, the total number of UPOV members is 44.

UPOV is gaining prominence as a legislative model for Plant Breeders' Rights because Article 27.3(b) of TRIPs obligates WTO members to adopt patents and/or 'an effective *sui generis* system' for plant varieties. Although no such system has been defined, UPOV asserts that it is the 'only internationally recognized *sui generis* system for the protection of plant varieties'.[195] In addition, a number of influential bodies, including the WTO, are pushing for a narrowing of the *sui generis* option to the legislative model provided by UPOV.

At the June 1999 Congress of the International Association of Plant Breeders for the Protection of Plant Varieties (ASSINSEL), private and public plant breeders from 31 industrialized and developing countries representing over 1000 seed companies met to define their position on the protection of

States that are party to the International Convention for the Protection of New Varieties of Plants (status as of 1 July 1999)

- Party to the 1961 Act, revised 1972:
  Belgium and Spain.
- Party to 1978 Act:
  Argentina, Australia, Austria, Brazil, Bolivia, Canada, Chile, China, Colombia, Czech Republic, Ecuador, Finland, France, Hungary, Ireland, Italy, Kenya, Mexico, New Zealand, Norway, Paraguay, Poland, Portugal, Slovakia, South Africa, Switzerland, Trinidad and Tobago, Ukraine, Uruguay.
- Party to 1991 Act:
  Bulgaria, Denmark, Germany, Israel, Japan, Netherlands, Republic of Moldova, Russian Federation, Slovenia, Sweden, United Kingdom, United States of America.

See p. 111 *et seq.* for the Notes

intellectual property. ASSINSEL concluded that the type of protection for plant varieties varies according to the technical, legal and socioeconomic status of the country. Developing country members concluded that it was premature to develop protection of plant varieties through utility patents. The ASSINSEL Congress recommended that developing countries adopt a *sui generis* system based on the 1991 Act of the UPOV Convention.[196]

Many developing countries currently do not confer protection on plant varieties in any manner. While there is ample scope for national discretion in interpreting the *sui generis* option, deadlines for the implementation of TRIPs Article 27.3(b) are fast approaching. As a result, many developing countries are under considerable pressure to consider UPOV's *sui generis* model for IP protection of plant varieties.

Members of the Crucible Group do not agree on whether or not and to what extent UPOV 1991 restricts the right of farmers to save protected seed for their own use.

Some CSOs and some developing countries view the 1991 Act as a more stringent form of plant variety protection that dramatically strengthens the rights of commercial breeders while narrowing the rights of farmers. They believe that the 1991 Act is biased towards the interests of industrial breeders and does not adequately protect the rights of farmers and community innovators.[197] UPOV 1991 significantly extends the rights of breeders and the scope of protected material (see discussion below). The cumulative effect, insist some CSOs, is that plant variety protection laws based on UPOV 1991 increasingly resemble the 'industrial strength' protection afforded by patents.

UPOV asserts that strong intellectual property protection is necessary to ensure an acceptable return on research investment and to encourage further plant breeding research that is essential to meet the challenges of increasing food production in the coming years. Proponents of UPOV insist that critics are misinterpreting the 1991 Act. While UPOV 1991 clearly strengthens the breeder's position in specific ways, it also preserves the ability of member states to allow farmers to save and replant protected seed (this is described by UPOV as the farmers' exemption or as the 'farmers' privilege').

Why the sharp difference of opinion surrounding the interpretation of UPOV 1991, Article 5(1) of the 1978 Act is generally interpreted by governments and CSOs to allow farmers to save and replant protected seed for their own use (without payment of royalties). The critical clause does not mention farm-saved seed at all, it simply indicates that the authorization of the breeder is not required for the production and non-commercial marketing of protected material.[198] UPOV proponents insist there is really no difference between the 1978 Act and the 1991 Act of UPOV in this regard — as neither Act, they argue, mandates an exemption allowing farmers to fully use farm-saved seed.

The 1991 Act leaves each member state free to include the farmers' exemption in national legislation ('within reasonable limits and subject to

See p. 111 *et seq.* for the Notes

the safeguarding of the legitimate interests of the breeder').[199] Under the 1991 Act, the farmers' exemption explicitly becomes an *option* for national laws. Some interpret this to mean that the farmers' exemption is no longer an automatic feature of the international rules governing plant protection under UPOV. Without positive action by member states on behalf of farmers, some fear that the farmers' right will be lost or significantly restricted. Who will determine the 'legitimate interests' of breeders? Under pressure from commercial breeding interests, some CSOs are concerned that the rights of commercial breeders will take precedence over the rights of farmers in national laws.

The 1991 Act of UPOV is interpreted as allowing a mandatory provision for subsistence farmers to use farm-saved seed.[200] In those cases where subsistence farmers are farming for a non-commercial purpose — to produce food for their families, they would fall squarely within that exclusion. However, virtually all subsistence farmers market some portion of their harvest. If subsistence farmers are using protected varieties for commercial purposes, the exemption for farm-saved seed would not apply. Some CSOs believe that denying poor farmers the option of trading or selling planting material within their customary markets could restrict their role in maintaining genetic diversity and enhancing local plant breeding, and ultimately threaten food security. Proponents of UPOV point out that it is very unlikely that plant breeders would sue a subsistence farmer even if she markets a portion of her harvest.

The 1991 Act extends breeders' rights to the harvested product of the protected variety, if that variety has been used without authorization of the breeder (infringement of breeders' rights).[201] However, if the variety has been used with the authorization of the breeder (that is, if a royalty has been paid on the seed), the breeder cannot claim rights on the harvested material.

Some CSOs are critical of provisions extending the scope of the breeders' rights beyond the reproductive material to the harvested material. Under Article 14(1) of the 1991 Act, for example, explicit authorization of the breeder is required to sell, export or import propagating material of the protected variety, among other activities. If a country that is not a member of UPOV cultivates a UPOV-protected variety (without authorization of the breeder), then the breeder could prevent that country from exporting its harvest into UPOV territory. Without these provisions, assert UPOV proponents, commercial plant breeders operating in global commodity markets would be unable to protect their protected material from country to country. Some CSOs assert that the wider scope of these provisions will give seed companies too much control over the rights of farmers and the food system.

UPOV 1991 extends coverage to all plant genera and species. Under UPOV 1978 plant varieties could be protected under plant breeders' rights or by patent law; but countries could not allow both patents and plant variety rights for the same species of plant. The 1991 Act allows for double protection under both breeders' rights and patents, if countries so elect.

UPOV 1991 introduces the concept of 'essential derivation', which is designed to prevent the practice of cosmetic breeding. If a new plant variety differs from an older one by a minor modification — the insertion of a single gene, for example — the new variety is deemed 'essentially derived' from the older one. The concept of essential derivation aims to protect breeders from piracy. The functional and legal application of the term 'essentially derived' has yet to be implemented. It is not certain what minimum genetic distance will be required to distinguish between protected varieties, especially for minor crops.

Advocates of the concept of essential derivation say that it will foster the breeding and development of new and increasingly productive varieties; cosmetic breeding should be eliminated because it has the negative effects of reducing genetic diversity, increasing genetic vulnerability and makes no positive contribution to improved productivity. Advocates believe that the concept of essential derivation will protect farmers, as well as breeders, from piracy.

Some CSOs are concerned that the criteria for determining essential derivation will work in favour of corporate breeders and will put farmers and breeders in the South at a disadvantage. There is concern that the use of high-technology mechanisms (molecular marker profiles, pedigree distance data) to determine the level of genetic distance and innovative skill in plant breeding will discriminate against community plant breeders and give too much control to IP regimes.

Alternative forms of plant variety protection that may implement 27.3(b)

Although the TRIPs Agreement does not give any details on what elements an effective *sui generis* system would have to include, certain minimum requirements that such a system would have to fulfil may be drawn from the context of Article 27.3(b), the context of the Agreement as an integral part of the WTO Agreement and, finally, from the objectives of the TRIPs Agreement itself.

Since the TRIPs Agreement elaborates no further upon the term 'plant variety', member states might have to provide for the protection of plant varieties of *all* species and botanical genera. The *sui generis* system has to be an intellectual property right. It needs to comply with the basic principle of national treatment. Thus, members have to accord to the nationals of other member states treatment no less favorable than that which they accord to their own nationals with regard to the protection of plant varieties. Furthermore, any advantage, favor, privilege or immunity granted by a member to the nationals of any other country has to be accorded immediately and unconditionally to the nationals of all the other member states (most-favored-nation treatment).

Finally, the *sui generis* system has to be effective. While some argue that

See p. 111 *et seq.* for the Notes

the term 'effective' refers to a certain minimum level of protection, others argue that the *sui generis* system is effective if it provides for an enforcement procedure so as to permit *effective legal* action against any act of infringement of the *sui generis* right.

Thus, although the *sui generis* system has to comply with certain basic requirements, it allows countries to develop national plant variety protection laws that are distinct from the UPOV model. Alternative proposals being discussed at the moment include the proposal to provide for the protection of farmers' varieties, to protect varieties only if the country of origin of the breeding material is disclosed or to grant rights that are weaker than the rights granted to breeders under the 1978 or the 1991 UPOV Acts[202] (see Volume 2 for further discussion).

Many countries and civil society organizations are engaged in processes to elaborate national plant variety protection laws that also embrace Farmers' Rights, community rights and systems for managing genetic resource use and access. Some of these aim to construct mechanisms for the exchange and transfer of biological resources, and to remunerate local communities for their contributions. In most countries, these laws have not yet been adopted. The elaboration of plant variety protection laws that seek to incorporate the rights of farmers and local communities is evolving rapidly, and experience is very recent. A more complete discussion of community rights regimes and legislative options is found in Volume 2.

Genetic Resources Action International (GRAIN), a CSO based in Barcelona, has recently compiled a summary of non-UPOV *sui generis* plant variety protection laws under debate in developing countries.[203] The following are just a few examples:

- At the Organization of African Unity's (OAU) Summit in June 1998, African governments agreed to develop an 'African Common Position' to safeguard the sovereign right of member states, and to merge the interests of national plant breeders with the national concerns of farmers, and Africa's political commitment to Farmers' Rights.[204] The OAU has since developed the 'African Model Legislation for the Recognition and Protection of the Rights of Local Communities, Farmers and Breeders, and for the Regulation of Access of Genetic Resources'. The model legislation will be available in a final draft form in early 2000.
- The Zambian government has drafted a plant variety protection law that seeks to protect the innovations of local communities and indigenous peoples, in keeping with its obligations under the CBD. The draft is now undergoing a process of wide public consultation.
- India's draft plant variety protection act attempts to balance recognition for Plant Breeders' Rights and Farmers' Rights. It states: 'Nothing contained in this Act shall affect a farmer's traditional right to save, use, exchange, share or sell his farm produce of a variety protected under this Act except where a sale is for the purpose of reproduction

under a commercial marketing arrangement.' The draft PVP law also provides for communities to register collective rights.

- In Thailand, two bills have been drafted that aim to recognize traditional knowledge and the rights of local communities. The draft Plant Variety Protection Bill would combine recognition for the rights of plant breeders to their newly developed varieties with the protection of native varieties that have been conserved and developed by farmers and local communities. The draft Traditional Medicine Bill would recognize rights to traditional healers and medicinal genetic resources, based on the concept of collective rights. It includes the registration of traditional medicines and some form of benefit-sharing in case medical or scientific researchers make use of the protected knowledge.

- The Plant Varieties Act of Bangladesh, drafted by the National Committee on Plant Genetic Resources, recognizes community rights and Farmers' Rights, and proposes the establishment of a fund to support communities in the conservation and development of plant varieties. The draft law is undergoing a process of public debate.

- Costa Rica introduced a plant breeders' rights bill at the end of 1999 for the purpose of complying with TRIPs. Because of concerns raised by civil society organizations, the bill is undergoing a process of public review and consultation. However, Costa Rica's 'Biodiversity Law' was signed into law in May 1998. The Biodiversity Law aims to comply with the mandate of the CBD. [205] Under the term 'sui generis community intellectual rights', the law recognizes and expressly protects the practices and innovations of indigenous peoples and local communities related to the use of biodiversity components, and their associated knowledge. The National Seed Office and the Intellectual Property Office are obliged to consult with the National Commission for Biodiversity Management (CONAGBIO) on innovations which involve biodiversity components prior to granting IP protection. The law obliges CONAGBIO's Technical Office to reject any request for recognition of intellectual or industrial rights for biodiversity components or knowledge that is already recognized by community rights.

Some members of the Crucible Group ask whether or not it is possible for governments to pursue policies that genuinely stimulate innovative research and still to craft a responsible legislative answer to Article 27.3(b) of WTO's TRIPs chapter. The divide between those who see intellectual property as a cost-effective 'win-win' for society and for inventors, and those who see its exclusive monopoly provisions as antithetical to innovation and society's needs, is too great. In the light of this divide, the Crucible Group does not provide a definitive answer but lists legislative options (see Volume Two) along with possible implications of the proposed choice. The Group stresses that the discussion should not lead readers to conclude that Crucible considers any legislative option to be sufficient or desirable at this

See p. 111 *et seq.* for the Notes

time. Some Group members regard all of the options to be fundamentally flawed.

Could Plant Breeders' Rights (plant variety protection) include farmers' varieties?

This issue highlights the widely divergent views that exist on the desirability of intellectual property. It is controversial also because there is little agreement about what constitutes a variety, whether such a system can be enforced and the balance between the production and distribution of new ideas. Many agree that the innovative activities of farmers and their communities must be recognized and encouraged to ensure food security and improve agricultural productivity, to conserve and create agricultural biodiversity; less certain is whether a *sui generis* system that provides protection for farmers' varieties will help realize these objectives.

The discussion over the inclusion — or not — of informal IPRs for farmers' varieties raises questions regarding the potential to amend existing conventions. However, many observers would agree that existing legislation and conventions could be amended to better accommodate the special concerns of farmers and indigenous peoples. In Volume Two, we examine ways in which such amendments could be introduced into national law.

Viewpoint box: Should farmers' varieties receive intellectual property protection?

Well worth doing
A class of protectable subject matter can be drawn up with clear guidelines that define a plant variety. A broad or narrow definition of the term would be determined by the extent and type of protection to be conferred. By providing for a broad definition of a plant variety, narrow limits would then be set for the level of protection. At the very minimum, varieties would need to be sufficiently distinguishable and describable in order to be eligible for protection. Varieties that satisfy a stricter criterion would qualify for stronger and/or longer protection.

Possible but at what cost?
The intention is commendable. But the actual task of drawing up a class of protectable subject matter is formidable. Laws can be written but must be enforceable. How wide should the definition of a variety be, so as not to disenfranchise any farming community and still be an incentive to innovation and conservation? Should all species that contribute to food and agriculture be considered or should species be limited to those considered vital for food security? Would the implementation of such a system overburden the appropriate authorities?

Pointless
This system must strike the right balance between the creation of new varieties and their spread, two inherently conflicting objectives. Bigger incentives and stronger rights lead to more innovation. This would, however, lead to a narrow definition of a variety with the risk of not covering farmers' varieties. A wider definition but weaker rights would, at best, prevent misuse and misappropriation of farmers' varieties but not provide any incentive to ensure food security and improve agricultural productivity, neither would it help conserve and create agricultural biodiversity.

See p. 111 *et seq.* for the Notes

The right of farmers to save and exchange proprietary seed
There are fundamentally different views on the desired scope of the farmers' exemption (that is, the right of farmers to save and exchange proprietary planting material under IP laws). A *sui generis* system for protection of plant varieties may be different from the UPOV model with regard to the farmers' exemption. While Crucible Group members do not agree on the scope of the farmers' exemption, the Group believes that an element-by-element examination of the issues is helpful and can lead to progress.

One perspective argues that every farmer and farming community has the inalienable right to save and exchange any breeding material in any way they wish, including through commercial sale of IP-protected germplasm. Another view is that such an unrestricted use of IP material is a violation of their rights and a threat to the future of world food security. Still others are prepared to recognize the rights of certain groupings of farmers, possibly defined by economic status, land holdings, or their use of planting material, to utilize seed in ways that would be prohibited to other groups of farmers.

Some members of the Group believe that plant breeders with proprietary varieties are, in general, prepared to accept that farmers who customarily plant back harvested seed because they lack access to — or financial resources for — new seed every growing season should be allowed to continue to do so. Further, such farmers should be permitted to exchange planting material with their neighbors since this activity represents a traditional form of community plant breeding that could readily fall under the normal breeders' exemption. Further still, farmer/breeders could be granted the privilege of trading and even selling seed within their customary market area. Proprietary breeders stress that such activities would have to be consistent with traditional practices and not merely an evasion of the rights of proprietary plant breeders to control commercial exploitation of protected propagating material. It is recognized that, among poor or primarily subsistence farmers, denial of the option to trade or sell planting material within their customary market could deny farmers an important mechanism through which they are able to maintain genetic diversity and enhance local plant breeding.

Proprietary breeders, in the main, also recognize that many poor farmers and poor countries work within geographic and economic environments within which it is difficult for the commercial sector to provide full and consistent services. In the same way, breeders accept that, from time to time, national emergencies and other threats to food security may arise that require the suspension of their rights so that urgent human needs can be addressed. Such situations are adequately covered under normal intellectual property conventions and other international undertakings.

Proprietary breeders point out, however, that any suspension of rights that they regard to be appropriate and consistent with IP conventions must be monitored carefully and that they should have the opportunity to challenge and seek redress where they consider their rights to have been

See p. 111 *et seq.* for the Notes

abused. Further, breeders feel that much of the international furore over this issue has incorrectly pitted farmers as 'David' versus multinational enterprises as 'Goliath'. Even subsistence farmers on marginal lands in poor economies could benefit from the development of local entrepreneurship in both plant breeding and seed multiplication and distribution, through private or cooperative organizations, that might improperly be constrained if farmer plant-back goes unmonitored and opportunities for commercialization are restricted unduly. Governments should seek to encourage appropriate local entrepreneurship at every opportunity.

Others in the Crucible Group regard the right of farmers and farming communities to save and exchange, including sell, planting material as vital to food security and essential to the conservation and enhancement of plant genetic resources. Any effort to constrain this right should be scrupulously studied and challenged. Limitations to this right based upon land holdings, economic status, or purpose are not valid within the understanding of customary use or in recognition of the importance of this practice for the conservation and enhancement of diversity for future generations.

The 1999 review of Article 27.3(b)

A much-anticipated review of Article 27.3(b), the provision calling for patents or *sui generis* rights on plant varieties, was scheduled to take place in 1999. However, there is growing uncertainty about the nature of the review itself. When the TRIPs Council met in April 1999, there was disagreement about whether the exercise was merely a review of member states' efforts to implement Article 27.3(b), or if it should also include a review and renegotiation of the text.[206] Throughout 1999 the TRIPs Council collected information from industrialized and developing country members to determine the status of implementation of Article 27.3(b). Some industrialized countries, including the US and EU members, prefer a simple information-gathering exercise, rather than a renegotiation of the text. However, a number of developing countries are proposing to reopen the negotiations and amend the language of Article 27.3(b).

Final decisions on the scope of the review, and whether or not Article 27.3(b) is to be reopened and renegotiated, are still pending. Attention shifted away from the review of Article 27.3(b) (or at least away from the idea that it might occur in isolation) as member states prepared to launch a new round of trade negotiations at the WTO Ministerial Conference in Seattle, 29 November–4 December 1999. For a while, it looked as though the review might be dealt with simultaneously, along with a host of other issues. Given member states' failure to launch a new round of negotiations, they will once again be faced with deciding what to do about the review on its own.

Sharply contrasting proposals are emerging for the renegotiation of Article 27.3(b). The US and other industrialized countries may push for deletion of the entire subparagraph, thus eliminating exceptions from patentability.

See p. 111 *et seq.* for the Notes

WIPO, UPOV and some UPOV member countries suggest that UPOV 91 should be explicitly named as an (or the) 'effective' *sui generis* system. By contrast, some developing countries and CSOs are advocating for an expansion of what may be excluded from patentability under TRIPs to provide member states with the option of excluding all biological materials from patentability.[207] While the US did not anticipate or favor new negotiations on TRIPs at the Seattle Ministerial meeting, some developing countries submitted proposals to the WTO TRIPs Council for new negotiations.[208]

At the 20–22 October 1999 meeting of the WTO TRIPs Council, the US and India each submitted papers concerning Article 27.3(b).[209] The US favors the US-style patent-based model for protection of plant varieties, and warns that any *sui generis* model for plant variety protection that is not modeled on UPOV 1991 would need to be examined on a case-by-case basis. By contrast, India asserts that the TRIPs Agreement conflicts with the CBD, and that the two must be reconciled before they can be implemented at the national level. India's paper focused on the limitations of IP regimes in adequately addressing aspects of indigenous knowledge. India advised developing countries to consider different models of protection before rushing to implement *sui generis* systems.

Ultimately, consensus will be required on any amendments to Article 27.3(b). Given the fast-approaching deadlines and the contentious nature of plant and animal patenting, consensus may be difficult to reach.

Viewpoint box: Patenting life forms?

No life patenting
It is a corporate world: multinationals generally get what they want and governments do little to protect the interests of ordinary citizens. Living materials are not inventions. In the current scientific and commercial environment, any intellectual property for any form of living material will inevitably evolve into patent monopolies over every form of life. The line must be drawn as clearly as possible. TRIPs should forbid all patents on elements of life: as a minimum it must drop its requirements for protecting plant varieties and microorganisms.

Steady as she goes
The provisions of Article 27.3(b) of TRIPs have yet to lead to any of the disasters that some foretold. However, experience with the system has so far been short. Provisions for patenting plants and animals arouse strong emotions, and even supporters acknowledge that not all problems have yet been sorted out, even in developed countries. It is too early to make further extension compulsory.

Move with the times
Scientific and technological advances drove the economic progress of the 20th century. Intellectual property is as important now as physical property was in the past. IP in all areas is essential to fund further advances. Inevitably, the broadening of innovation to biological resources creates complication and uncertainty, but these are transitional problems quickly overcome. There is no logic to retreating into history. TRIPs must be strengthened to require strong patent protection for all new and inventive biological materials. All the exemptions of Article 27.3(b) should be eliminated.

See p. 111 *et seq.* for the Notes

The WTO's TRIPs review has provided a sharp focus for the international debate surrounding intellectual property claims involving biological material. Among many farm and indigenous organizations, opposition to 'life patenting' is mounting. In between, lies a group that believes it is too early to extend obligations to patent life-forms.

Crucible Recommendation 15

Amending TRIPs?
The Crucible Group reiterates its 1994 recommendation that developing country signatories to the WTO Agreement take full advantage of the opportunities and flexibilities available in the TRIPs chapter, with respect to plant varieties, to exercise national sovereignty and choice and to develop the most appropriate legislative tools for the furtherance of agricultural development in their countries. Governments should bear in mind that it is not necessary to have membership in any intergovernmental convention in order to have an effective *sui generis* law. As governments consider the adoption of legislation, the Group recommends that:
- the option to refuse patent protection for plants and animals should be maintained. Further, the option to protect plant varieties by effective *sui generis* legislation for plant variety protection given by Article 27.3(b) should be retained;
- the TRIPs Council should recognize that capacity-building in least developed countries has to take place before legislation can sensibly be introduced, and grant appropriate extensions to comply with treaty obligations.

The CGIAR and intellectual property
A rapidly changing IP environment and increasing privatization of agricultural research has forced the CGIAR to develop policies and procedures on IP over the past decade. The process has been complicated by the fact that the CGIAR System has no legal status, and its members often represent opposing sides of the highly politicized IP debate. In addition, there are at least 14 'policy-making' bodies within the CGIAR. After years of discussion and debate by numerous committees, the CGIAR System is still in the process of developing a coherent, comprehensive policy on IP. Given the rapidly changing international policy environment and ongoing debate in many international fora, the CGIAR decided in 1996 to endorse 'Guiding Principles for the CGIAR Centres on Intellectual Property and Genetic Resources' as an 'interim working document that will be continually reviewed and revised'. Among other principles: 'The Centres will not claim legal ownership nor apply intellectual property protection to the germplasm they hold in trust, and will require recipients of the germplasm to observe the same conditions, in accordance with the agreements signed with FAO.'

The use of proprietary materials and technologies is an increasingly complex policy challenge for the CGIAR. In 1998–99 the CGIAR Panel on Proprietary Science and Technology was charged with distilling the complex

See p. 111 *et seq.* for the Notes

policy issues of intellectual property and its role in the future of CGIAR. The FAO/CGIAR Agreement obliges centres to exclude IP protection over 'in-trust' germplasm. That much is clear. But what about IP protection for technologies and materials developed by CG scientists? Should CGIAR centres seek proprietary rights, and, if so, under what circumstances? A survey conducted by the International Service for National Agricultural Research (ISNAR) in late 1997 found that IARC scientists are routinely using proprietary technologies (i.e. selectable markers, gene promoters, transformation systems, etc.) in biotechnology research.[210] What are the

Viewpoint box: CGIAR and intellectual property

Pursuing IP
Proprietary science has profoundly changed publicly-funded agricultural research, and access to and exchange of genetic resources. More than 70% of cutting edge agricultural biotechnology is produced and controlled by the private sector, and those technologies are central to the goal of increasing food production. In order to gain access to relevant technologies, the IARCs need to develop partnerships with the private sector and participate proactively in global IP regimes. The IARCs need 'bargaining chips' so that they can begin to trade patented technologies with the private sector. New technologies that emerge from IARCs must be protected to ensure that they can be used for the benefit of developing country NARS and poor farmers. IP agreements are the key to facilitate commercialization and technology transfer. In the future, they may even become a source of revenue for CGIAR (though this is much less important than ensuring access to and use of the technology).

Defensive use of IP
Most of CGIAR's work is achievable without a major shift of emphasis toward the use of proprietary science. However, the CGIAR must seek and defend IP for innovations in those cases where it is necessary to prevent their appropriation by private claimants, and to ensure that they remain in the public domain. The important thing is not to lose sight of the primary mission: to serve the poor. In light of declining budgets, CGIAR must carefully weigh the substantial costs of increasing its capacity to manage IP against competing needs. CGIAR should take a cautious role in advocating IP for biological materials. It would be useful for CGIAR to campaign for a clearer definition of the research exemption under patent law, for example. All of this can be done without compromising CGIAR's mission to produce international public goods.

Rejecting IP
CGIAR should avoid proprietary science because it has little or no relevance to its mission of alleviating hunger and poverty, and it will distort the CGIAR's mandate to serve poor farmers. IP will become a barrier to accomplishing the CGIAR's mission. IP claims on biodiversity often appropriate the resources and knowledge of farmers and indigenous people. Proprietary science is already inhibiting the open exchange of genetic resources and knowledge, and it has a chilling effect on agricultural research and innovation. Filing patents and defending them is expensive, and it will take already scarce resources away from agricultural research. Instead of forging commercial partnerships with the biotech industry for capital-intensive technologies that promote industrial monoculture, IARCs should instead focus on building partnerships with resource-poor farmers, and make use of indigenous knowledge in local agro-ecological systems. CGIAR should take a clear position against all IP agreements.

See p. 111 *et seq.* for the Notes

legal implications of using someone else's proprietary inputs in CGIAR research? Are IARC scientists free to disseminate results derived from proprietary science? Does the use and development of proprietary science within the CGIAR distort or advance its mission to promote food security for the poor? These are among the complex questions facing the CGIAR as it struggles to define its role as the world's premiere public sector agricultural research body in an era of rapidly evolving IP regimes, proprietary science and declining research budgets.

Despite the lack of a clear policy mandate, several IARCs are pursuing IP protection on CG-related research. The International Maize and Wheat Improvement Centre (CIMMYT), for example, is named as a co-inventor on a 1997 patent relating to apomictic maize. CIMMYT describes the patent as a defensive move to ensure that it can make apomictic maize freely available for resource-poor farmers of the developing world.[211] Other IARCs have applied (or are expected to file) for patents on a new animal vaccine and on an improved crop line for virus resistance.[212]

In 1998 the CGIAR expanded its position on IP. It confirmed a set of guidelines on genetic resources and IP, which includes the Trust Agreement on genetic resources with FAO. It also established a special unit at ISNAR to provide legal counsel on IP to IARCs. In 1999, the CGIAR Consultative Council asked the Centre Board Chair's Committee and the Centre Director's Committee to look into the possible next steps. Is there a need to update the system-wide IP audit? Should each IARC manage its own IP agenda, or should a centralized body manage IP within the CGIAR? The CGIAR has rejected a proposal to fully centralize ownership of IARC-protected technology through a new legal entity. Another possibility under discussion is the establishment of a wholly-owned subsidiary under full IARC control that would manage IP for all IARCs concerned.

See p. 111 *et seq.* for the Notes

Other developments

Intellectual property regimes are changing and evolving rapidly. In the following pages Crucible briefly reviews new and recent developments concerning IP. Despite concerted efforts to achieve harmony and consistency across national and regional borders, IP as it applies to life forms remains steeped in a climate of controversy and uncertainty.

European Parliament approves patent directive

In an effort to harmonize rules in the European Union, the European Parliament gave final approval to a controversial biotechnology 'patent directive' in May 1998. The European Directive on the Legal Protection of Biotechnological Inventions aims to harmonize national legislation on the patenting of genetic material within the EU. The Directive creates, for the first time in European history, an explicit legal right to obtain patents for higher organisms, such as plants and animals.[213] However, it does not create a European-wide patent,[214] nor is it binding for the European Patent Office.[215] The new law came into force on 30 July 1998. Member states have two years to implement its provisions.

For one decade prior to its approval, the patent directive was a lightning rod for vigorous debate throughout Europe on the ethics and morality of biotechnology and life patenting. In 1995, European civil society organizations opposed to life patenting lobbied strenuously and effectively to defeat an earlier draft of the patent directive. Despite adoption of the EU patent directive, biotech patenting remains controversial in Europe,[216] and the EU patent directive now faces legal challenges.

In October 1998 the Dutch government filed a nullity suit at the European Court of Justice against the EU's patent directive. The Italian government joined the Dutch government's challenge in early 1999. The new German government has also expressed concerns about the directive, and suggested it may not be able to pass the necessary legislation to enact it into German law.[217]

The EU patent directive clarifies the scope of patentability for biotech-related inventions within 15 EU member states. It allows for the patenting of transgenic plants and animals, provided that they meet standard criteria for patentability. In a nod to ethical concerns, the directive outlaws transgenic animal patents on inventions 'likely to cause [animal] suffering without any substantial medical benefit to man or animal'. The Directive also says that plant and animal patent applications should specify the geographic origin of patented material. In the case of human materials, the person from whom the genetic material was taken must have had the opportunity of giving free

See p. 111 *et seq.* for the Notes

and informed consent thereto 'in accordance with national law'. However, the directive provides no sanction for breach of these requirements.[218]

Humans and human embryos are excluded from patentability, but Article 5 of the directive recognizes that human genes 'isolated from the human body or otherwise produced by means of a technical process' can be patented, 'even if the structure of that element is identical to that of a natural element'. The directive also clarifies the EU position on the patenting of partial gene sequences, by requiring that the function or industrial application of a gene sequence be disclosed in all patent applications.

The patent directive allows for the so-called farmers' exemption. Small-scale farmers may freely use farm-saved seeds of specified plant varieties for propagation or multiplication on their own farm. Large-scale farmers are required to pay royalties on farm-saved seed.

The Directive does not include a provision for breeders to freely use plant varieties including patented biotechnological inventions as the initial source for creating other new varieties — the so-called breeder's exemption. As a result, breeders are left with uncertainty about whether or not and under what conditions they may use a plant variety containing patented traits for breeding purposes.[219]

From the viewpoint of those opposed to the patenting of life forms, approval of the patent directive opened the floodgates to the 'industrial commodification' of all life forms, eliminating Europe's last symbolic barrier to legal resistance. For proponents in government and industry, the patent directive offers much-needed clarification in a controversial area of law and provides a 'sensible compromise' between the views of the biotech industry and the ethical concerns that are surrounding the direction of genetic research.[220]

US legal challenge on validity of plant utility patents

In January 2000, a federal appeals court affirmed the patentability of plants and seeds in the United States. The ruling was a victory for Pioneer Hi-Bred (Dupont) because it affirms that the US Patent and Trademark Office has the authority to grant patents on sexually reproducing plants. The Federal circuit court ruled that plant breeders have their choice of intellectual property protection — either through utility patents or through the US Plant Variety Protection Act; and the court specifically held that the two acts were consistent with each other.[221]

The case concerned Pioneer Hi-Bred suing a seed merchant for infringing a patent by buying and re-selling 600 bags of Pioneer Hi-Bred's proprietary maize seed. The seed merchant argued that the case should be dismissed on the grounds that utility (industrial-type) patents on plant varieties are illegal in the first place. Ultimately, the court did not accept this argument.

See p. 111 *et seq.* for the Notes

Bilateral/unilateral action to protect IP

The establishment of the WTO is supposed to prevent members from unilaterally adopting trade measures to enforce trade goals. However, the US continues to initiate strong unilateral and bilateral measures, such as punitive sanctions against trading partners under the 'Special 301' annual review process.[222] Under this process, the US government (in close consultation with industry groups) examines the record of intellectual property protection in more than 70 countries and publicizes annually the names of trading partners that it believes are failing to provide adequate and effective protection to intellectual property. The US government's retaliatory actions fall into different categories, ranging from a warning to imposing trade sanctions.

In some cases, the US is pressing developing countries to accept stronger IP protection than required under the WTO TRIPs Agreement. For example, the US complained that Argentina's new patent law delayed extension of patents to pharmaceuticals until 2000. Under TRIPs, developing countries are not obligated to phase in patent protection of new product types until ten years after the Agreement entered into force — well beyond the year 2000.

Special 301 is considered by many, from different sides of the trade debate, to be counter to the aims of the WTO, and a breach of its rules which require that any trade sanctions should be preceded by a Dispute Settlement Panel decision and specific approval of retaliatory measures.

Multilateral agreement on investment

The Multilateral Agreement on Investment (MAI) is an international economic agreement that was negotiated within the framework of the Organization for Economic Cooperation and Development (OECD) between 1995 and 1998. The MAI aims to reduce obstacles and inefficiencies to overseas investments, with the goal of making it easier for corporations to move their investments — both capital and production facilities — across international borders. Citing concerns about national sovereignty, lack of labor and environmental protections and issues of corporate accountability, some LDCs and CSOs vigorously protested the lack of transparency in MAI negotiations, arguing that governments would be abdicating regulatory power to multinational corporations. After France withdrew from the MAI negotiations, the OECD suspended negotiations in December 1998. However, similar agreements to liberalize the movement of foreign investment are currently being promoted by the Transatlantic Economic Partnership. Some parties predicted that similar issues would be included in the Millennium Round of trade negotiations. Obviously, the failure of the member states meeting in Seattle to launch a new trade round (for the time being at least)

See p. 111 *et seq.* for the Notes

frustrated the efforts of those who would have liked to have seen them included.

Trilateral world patent system?

Patent offices in Japan, Europe and the United States handle approximately 80% of all patent applications worldwide. In November 1997 the heads of government patent offices from the US, the European Union and Japan met in Kyoto and agreed on steps to integrate their respective patent examination systems into a global network.[223]

The officials meeting in Kyoto agreed to establish a computerized 'Trilateral Patent Network' which will allow the patent offices to exchange technological and administrative data on technologies that are the subject of patent applications. The government officials attending the trilateral meeting said that efforts to integrate the three government patent offices may eventually lead to a patent being recognized and protected in all three areas simultaneously.

See p. 111 *et seq.* for the Notes

Concluding remarks*

This concludes Volume 1 of the Crucible II Project's reports. Even as this edition is being type-set for publication, significant developments are being made in biological sciences and in biotechnology. Likewise, the policy environment is rapidly changing. Government delegates and activist observers are gearing up for important international meetings such as the Ad-Hoc, Open-Ended Intercessional Meeting on Article 8(j) and indigenous and local knowledge in Seville in March 2000; COP V of the CBD in Nairobi in May 2000; the Global Forum on Agricultural Research and the CGIAR Mid-term Meeting in Dresden in May 2000, to mention a few. National governments are engaged in domestic policy making exercises. Peru for example, appears to be on the verge of introducing legislation regulating access and creating forms of *sui generis* intellectual property rights for traditional knowledge. Kenyan officials are working on national access legislation. Malaysia is working on plant variety protection legislation. It is quite likely that every country in the world is engaged in building up laws and policies that effect the way in which genetic resources will be regulated in that country, and by extension, internationally. Community groups, civil society organizations and indigenous peoples organizations are working hard on the local, national and international levels of policy making to make sure their voices are being heard. There are innumerable examples of such initiatives being undertaken while this book goes to print. In Canada, the British Columbia Council of Indian Chiefs is sponsoring a conference in mid-February, 2000 on the protection of indigenous peoples' knowledge.

It is a challenge to simply keep informed of all of the relevant developments in science, technology, policy, politics, shifts in popular opinion, and law in the field of genetic resources. It is even more difficult to try to make a constructive contribution to the way in which all of the various actors interrelate. Our intention in engaging in this Crucible Round was to work with as many different players with as wide a range of opinions as we could possibly assemble in one project to try to work through the morass of issues that constitute what we now refer to as the field of genetic resources. In doing so, we hope we have lent transparency to the variety of policy options which are available to policymakers and advocates in both domestic and international fora

The pace of change in the field is accelerating. We hope that this Volume (and Volume 2 which follows) will contribute to raising the level of common understanding of the way issues in this field relate to one another, and ultimately, move towards the creation of mutually satisfying policies.

---

* These remarks were added by the management committee immediately before the manuscript was sent to the printer.

See p. 111 *et seq.* for the Notes

Abbreviations

ACTS	African Centre for Technology Studies
ASSINSEL	Association Internationale des Séléctionneurs pour la Protection des Obtentions Végétales (International Association of Plant Breeders for the Protection of Plant Varieties)
CBD	Convention on Biological Diversity
CBDC	Community Biodiversity and Development Conservation Program
CGIAR	Consultative Group on International Agricultural Research (also called the CG System)
CGRFA	Commission on Genetic Resources for Food and Agriculture
CIMMYT	Centro Internacional de la Mejoramiento de Maiz y Trigo (International Maize and Wheat Improvement Centre)
CIP	Centro Internacional de la Papa (International Potato Centre)
COP	Conference of the Parties (to the Convention on Biological Diversity)
CSO	civil society organization
EST	expressed sequence tag
FAO	Food and Agriculture Organization of the United Nations
GATT	General Agreement on Tariffs and Trade
GMO	genetically modified organism
GPA	Global Plan of Action
GRFA	genetic resources for food and agriculture
GURT	genetic use restriction technology
HAC	human artificial chromosome
IDRC	International Development Research Centre
IP	intellectual property
IPCC	Intergovernmental Panel on Climate Change
IPGRI	International Plant Genetic Resources Institute (former IBPGR)
IPRs	Intellectual Property Rights
ISAAA	International Service for the Acquisition of Agri-biotech Applications
ISNAR	International Service for National Agricultural Research
IUCN	The World Conservation Union
LMO	living modified organism
NAFTA	North American Free Trade Agreement
OECD	Organisation for Economic Co-operation and Development
PBRs	Plant Breeders' Rights
PGRFA	Plant Genetic Resources for Food and Agriculture
PPB	participatory plant breeding
RAFI	Rural Advancement Foundation International
SINGER	CGIAR's System-wide Information Network for Gentic Resources
SNP	single nucleotide polymorphism
TAC	Technical Advisory Committee
TRIPs	Trade-Related Aspects of Intellectual Property Rights
UNCED	United Nations Conference on Environment and Development
UNDP	United Nations Development Programme
UNESCO	United Nations Educational, Scientific and Cultural Organization
WIPO	World Intellectual Property Organization
WTO	World Trade Organization

Notes

1. The CBD defines 'biological diversity' as the variability among living organisms from all sources including, *inter alia*, terrestrial, marine and other aquatic ecosystems and the ecological complexes of which they are part; this includes diversity within species, between species and of ecosystems.
2. United Nations Human Development Report, Oxford University Press, 1999, p. 70.
3. Intellectual property (IP) is often used as a collective name for rights such as patents, trademarks, trade secrets, copyrights, plant breeders' rights, etc. IP refers to private rights granted by a state authority to IP 'owners' for a specified time period, so that they can control whether or not, and under what circumstances, others can use their ideas or innovations.
4. In this document we use the term 'civil society organization' (CSO) instead of 'non-governmental organization' (NGO). The term refers to an association formed for collective purposes primarily outside of the state and marketplace. CSOs are active in shaping democratic and developmental goals at the national and/or international level. The term 'civil society' has a long and complex history in political philosophy. For a detailed discussion, see Van Rooy, Alison, *Civil Society and the Aid Industry: The Politics and Promise*, Earthscan, London, with the North-South Institute, Ottawa, 1998.
5. Not all members of the Crucible Group agree on whether or not global climate change poses a serious threat.
6. World Bank, World Development Report, *Knowledge for Development*, New York, Oxford University Press, 1999, p. 16.
7. *Ibid.*, p. 27.
8. Serageldin, Ismail, and Joan Martin-Brown (eds), 'Ethics and values: a global perspective: proceedings of an associated event of the fifth annual World Bank Conference on Environmentally and Socially Sustainable Development', World Bank, 1998, p. 46.
9. World Bank, *op cit.*
10. *Ibid.*, p. 34.
11. *Ibid.*, pp. 34-35.
12. *Ibid.*, p. 34.
13. *Ibid.*, p. 35.
14. The International Undertaking is in the process of being revised in harmony with the CBD. Farmer's Rights is one of the outstanding issues to be resolved. See detailed discussion in 'Outstanding Issues' chapter.
15. United Nations Human Development Report 1999, Oxford University Press, p. 25.
16. *Ibid.* p. 31.
17. FAO Press Release 98/69, 'FAO Releases Annual State Of Food And Agriculture Report Showing Worldwide Number Of Hungry People Rising Slightly Warns Of Slower Economic Growth In Most Developing Countries', FAO, Rome, 26 November 1998.
18. United Nations Human Development Report 1999, p. 28.
19. United Nations Population Fund, The State of the World's Population, 1996, p.1.
20. See, for example: Serageldin, Ismail, 'Biotechnology and Food Security in the 21st Century', *Science*, 16 July 1999; Rosset, Peter, 'Why Genetically Altered Food Won't Conquer Hunger', *New York Times*, 1 September 1999.
21. FAO, *Report on the State of the World's Plant Genetic Resources for Food and Agriculture*, 1996, p. 13, and IPGRI, *Diversity for Development*, 1999, p. 1. According to FAO, plant genetic resources for food and agriculture comprise the diversity of genetic material contained in traditional varieties and modern cultivars, as well as crop wild relatives and other wild plant species that can be used now or in the future for food and agriculture.
22. FAO Press Release, 'New FAO World Watch List for Domestic Animal Diversity

Warns: Up to 1,500 Breeds are at Risk of Extinction', 5 December 1995.

23. Estimate of 29 hectares per minute comes from Consultative Group on International Agricultural Research (CGIAR) press release, 'Poor Farmers could Destroy Half of Remaining Tropical Forest', 4 August 1996.

24. FAO Fisheries Department, 'The State of World Fisheries and Aquaculture', Rome, 1995, p. 8.

25. CGIAR, Statement of the CGIAR to the 4th meeting of the Conference of Parties to the Convention on Biological Diversity, Bratislava, Slovakia, May 1998.

26. Bryant, D., and L. Burke, *Reefs at Risk: A Map-Based Indicator of Threats to the World's Coral Reefs*, World Resources Institute, 1998.

27. Walter, K.S., and H.J. Gillett (eds), '1997 IUCN Red List of Threatened Plants. Compiled by the World Conservation Monitoring Centre', IUCN/The World Conservation Union, Gland, Switzerland, and Cambridge, UK, 1998. The authors note that the study relies on data found primarily in industrialized countries, and thus underestimates the true global situation.

28. Edwards, Rob, 'Save our pathogens', *New Scientist*, 22 August 1998, p. 5.

29. FAO Press Release 98/69, 'FAO Releases Annual State of Food and Agriculture Report Showing Worldwide Number of Hungry People Rising Slightly; Warns of Slower Economic Growth in Most Developing Countries', FAO, Rome, 26 November 1998.

30. Maffi, Luisa, 'Linguistic and Biological Diversity: The Inextricable Link', paper presented at the International Conference 'Diversity as a Resource: Relations between Cultural Diversity and Environment-Oriented Society', Rome, 2–6 March 1998.

31. See for example the discussion of the different names for the plant 'croton' among the different ethnic and linguistic groups in E.N. Meza & M. Pariona, 'Nombres Aborigenes Peruanos de las especies de Croton que producen el latex denominado 'Sangre de Grado'', in E.N. Meza, (ed.), *Desarrollando Nuestra Diversidad Biocultural: 'Sangro de Grado' y el Reto du su Produccion Sustentable en el Peru.*, Universidad Nacional Mayor de San Marcos Fondo Editorial, Lima, 1999.

32. Maffi, Luisa, *op. cit.*

33. *Ibid.*

34. Luisa Maffi cites R. Bernard in her paper, *ibid.*

35. According to Louise Sperling of CIAT, there are two general approaches to participatory plant breeding: formal-led and farmer-led. In farmer-led PPB farmers take part in crop breeding and seed supply activities initiated by agricultural scientists from research and development organizations. Formal-led PPB programs may aim to reshape formal breeding approaches (e.g. the type of germplasm used—local or exotic; the way testing is done—centralized or de-centralized. Formal-led PPB may also combine the goal of biodiversity enhancement with the more conventional goal of increasing production. With farmer-led PPB, external agents such as trained researchers, development officers and paraprofessionals support farmers' own systems of crop development. The objective of building farmers' skills to more effectively control and shape plant breeding becomes as important as identifying locally adapted and diverse varieties.

36. McGuire, S., G. Manicad, and L. Sperling, 'Working Document 2. Technical and Institutional Issues in Participatory Plant Breeding: Done from the Perspective of Farmer Plant Breeding', PRGA Program, Cali, Colombia, 1999. See also, Weltzein, E., M. Smith, L. Meitzner and L. Sperling, Working Document 3, Technical and Institutional Issues in Formal-Led Participatory Plant Breeding. A Global Analysis of Issues, Results and Current Experience, PRGA Program, Cali, Colombia, 1999.

37. The working group is part of the CGIAR System-wide Program on Participatory Research and Gender Analysis for Technical Development and Institutional Innovation.

38. Dr Robert Watson, Chair of the Intergovernmental Panel on Climate Change provided these findings at CGIAR's Mid-Term Meeting in Brasilia, July 1998. Source: *CGIAR News*, 'Climate Change Expert Speaks at MTM98', August 1998, p. 14.

39. Heywood, V.H., Executive Editor, 'Global Biodiversity Assessment', published for the United Nations Environment Program by Cambridge University Press, 1995, p. 321.

40. Watson, Robert, Chair of the Intergovernmental Panel on Climate Change. Remarks made during his presentation to the Mid-Term Meeting of the CGIAR in July 1998 in Brasilia. This data is taken from reprints of Dr Watsons slide presentation.

41. *Ibid.*

42. *Ibid.*

43. See the Technical Advisory Committee's Progress Report to the CGIAR, 'Climate Change and the CGIAR', October 1999. This document was circulated at International Centre's Week, October 1999.

44. Petit, M.J., The Emergence of a Global Agricultural Research System. ESDAR Special Report No. 1, The World Bank, Washington, D.C., 1996.

45. *Ibid.*

46. Alston, J.M., P.G. Pardey and J. Roseboom, 'Financing Agricultural Research: International Investment Patterns and Policy Perspectives', *World Development*, Vol. 26, No. 6, 1998, pp. 1057–1071.

47. *Ibid.*

48. *Ibid.*

49. *Ibid.*

50. Erbisch, F.H., and K.M. Maredia, *Intellectual Property Rights in Agricultural Biotechnology*, CAB International, Wallingford, 1998.

51. Brady, N.C., 'Quo Vadis International Agricultural Research', pp. 15–26. In C. Bonte-Friedheim and K. Sheridan (eds), *The Globalization of Science—The Place of Agricultural Research*, ISNAR, The Hague, 1998.

52. Anonymous, UC Berkeley and Novartis, 'An Unprecedented Agreement', *Global Issues in Agricultural Research*, Vol. 1, No. 3, 25 January 1999, p. 5.

53. Serageldin, I., *op. cit.*

54. Speech by Gordon Conway entitled 'The Rockefeller Foundation and Plant Biotechnology'. The speech was made to the Monsanto board of directors on 24 June 1999 in Washington, DC.

55. Personal communication with Richard Peterson, Securities Data Company. See also 'M & A Activity Hits Historic Levels', Securities Data Press Release, 5 January 1998.

56. Securities Data, Inc.

57. UNCTAD Press Release, 'Continued Upswing of Global FDI in 1996', 10 July 1997.

58. *Agrow,* No. 335, 27 August 1999, and RAFI, *RAFI News Release*, 3 September 1999, 'World Seed Conference: Shrinking Club of Industry Giants Gather for Wake or Pep Rally?', on the internet: http://www.rafi.org

59. The estimated market share of the top 10 seed companies depends on the estimated value of the commercial seed trade worldwide. ASSINSEL estimates the value of the commercial market for 'seed and planting material' at US$22.8 billion for 44 countries, and the total commercial world market at US$30 billion. Seed statistics from ASSINSEL (International Association of Plant Breeders for the Protection of Plant Varieties), are available on the internet: http://www.worldseed.org/~assinsel/stat.htm. RAFI estimates the value of the commercial seed trade at US$23 billion. See: 'Seed Industry Giants: Who Owns Whom' and *RAFI News Release,* 3 September 1999, on the internet: http://www.rafi.org

60. RAFI, 'The Gene Giants: Masters of the Universe?', based on material provided by Fountain Agricounsel, LLC, *RAFI Communiqué*, March/April 1999.

61. *Ibid.*

62. Examples of major life industry firms include: Novartis, Monsanto, DuPont, Hoechst and AstraZeneca.

63. Bratic, W., P. McLane and R. Sterne, 'Business Discovers the Value of Patents', *Managing Intellectual Property*, September 1998.

64. Dr Rupp was quoted on Hoechst web site, 1996: http://www.hoechst.com/press-e/13096e3.htm

65. Novartis, 'We Are Novartis', Novartis Communication, Basel, Switzerland, March 1997.

66. Grooms, Lynn, 'With Merger Completed, Harris Moran Focuses on Future', *Seed & Crops Digest*, January 1999.

67. Hayenga, M., 'Structural Change in the Biotech Seed and Chemical Industrial Complex', *AgBioForum*, Vol. 1, No. 2, Fall 1998.

68. *Ibid.*

69. Technology markets can be segmented in two main ways: by product or by territory. Both depend on intellectual property rights, in varying degrees. One example of market segmentation is a deal brokered by ISAAA between Monsanto and the Mexican government to license Monsanto's antiviral GM technology for use in potatoes. The technology was made available—free of charge—to Mexico for use in two locally grown potato varieties. There is no export market for these varieties, and little trade in them in Mexico, so they would not compete in markets in which Monsanto could charge an economic price. Here the segmentation was probably not dependent primarily on IP rights, so much as the difficulty of transferring the technology out of the varieties in which they were sold. Similar arrangements can be visualised in other crops, particularly crops which are of more importance for subsistence than for trade.

70. James, C., 'Global Review of Commercialized Transgenic Crops: 1998', ISAAA Briefs, No. 8. ISAAA, Ithaca, NY, 1998.

71. *Ibid.*

72. Bernard Le Buanec, Secretary-General of the International Seed Trade Federation, speaking on 16 January 1998. On the internet at: http://sci.mond.org/pubs.html, News Direct Alert C-127, 1/20/98.

73. James, C., *op. cit.*

74. Anonymous, 'Food for Thought', *The Economist*, 19 June 1999.

75. The reasons underlying the lack of agreement regarding the scope, terms, and applicability of the precautionary principle vary from issue to issue, but can be said to derive from certain basic themes. They include: the level of risk that triggers the need for precautionary measures; the nature and amount of scientific knowledge that must exist or not exist to implicate the principle; the degree of precaution that is warranted once the principle is deemed to apply; and the burden on economic efficiency that is justified by the avoidance of risk under the principle. Perhaps the greatest source of uncertainty regarding the definition of the precautionary principle is whether it (a) requires no further action with regard to a particular activity unless the proponents of that activity can demonstrate the complete absence of risk or (b) requires that the activity in question be halted only where a discernible risk—even a rather small or remote risk—can be demonstrated. It is this fundamental difference between demonstrating the complete absence of risk versus the existence of some risk that seems to be the fundamental divide separating interested observers.

76. Anonymous, PR Newswire, Washington, DC, 10 June 1999.

77. Saegusa, Asako, *Nature*, 24 June 1999.

78. Minutes of the Council Meeting, 24–25 June 1999. Annex: 'Declaration Regarding the Proposal to Amend Directive 90/220/EEC on Genetically Modified Organisms'. Declaration by the Council and the Commission.

79. Anonymous, 'A Golden Bowl of Rice', *Nature Biotechnology*, Vol. 17, September 1999, p. 831.

80. *Ibid.*

81. Monsanto Press Release, 'Monsanto Releases Seed Piracy Case Settlement Details', 29 September 1998.

82. Marshall, Eliot, 'Whose DNA is it, anyway?', *Science*, Vol. 278, 24 October 1997, p. 565.

83. Moran, N., 'Roche to Pay DeCode $200 Million for Disease Gene Discovery', *BioWorld Today*, 3 February 1998.

84. Marshal, Eliot, 'Tapping Iceland's DNA', *Science*, Vol. 278, 24 October 1997, p. 566.

85. 'DeCode Deferred', letter to the editor in *Nature Biotechnology*, Volume 16, June 1998, p. 496. See also, Andrews, Lori, and Dorothy Nelkin, 'Whose body is it anyway', *The Lancet*, 3 January 1998.

86. Schwartz, John, 'With Gene Plan, Iceland Dives Into a Controversy: Sale of Citizens' Genetic Code Pits Privacy Issues vs. Science', *International Herald Tribune*, 13 January 1999.

87. See website established by Mannvernd, advocacy group formed to promote ethical standards in medical research, on the internet at: http://www.simnet.is/mannvernd/english/index.html

88. Cohen, P., 'Hold the Champagne', *New Scientist*, 14 November 1998, p. 6.

89. Mahnaimi, U., and M. Colvin, 'Israel Planning Ethnic Bomb as Saddam Caves In', *Times of London*, 15 November 1998.

90. Kutukdjian, G., 'The need for bioethics is universal', *Biotechnology and Development Monitor*, No. 31, June 1997, p. 24.

91. Monsanto, Annual Report, 1997.

92. Anonymous, quoting William Haseltine, CEO of Human Genome Sciences, Inc., 'Company Hopes Gene Trawl Will Bring in Big Cash Catch', Reuters, 16 July 1999.

93. Cohen, P., 'Who Owns the Clones', *New Scientist*, 19 September 1998, p. 4.

94. Pennisi, E., 'Cloned Mice Provide Company for Dolly', *Science*, 24 July 1998, Vol. 281, pp. 495–97.

95. Cohen, P., 'Who Owns the Clones?', *op cit.*

96. Lemonick, M., 'Dolly, You're History', *Time*, 3 August 1998, p. 64.

97. Brower, V., 'Cloning Improvements Suggested', *Nature Biotechnology*, Vol. 16, September 1998, p. 809.

98. Food and Agriculture Organization of the United Nations, 'Working Group on the Implications of Developments in Biotechnology for Conservation of Animal Genetic Resources—Reversible DNA Quiescence and Somatic Cloning, Report of a Workshop, 26–28 November, 1997', Rome, 1998.

99. Anonymous, 'Cloning Saves Endangered Breed', *AgBiotech Reporter*, September 1998, p. 19.

100. Schuman, M., M. Waldholz and R. Langreth, 'Korean Experiment Fuels Cloning Debate', *Wall Street Journal*, 21 December 1998.

101. *Ibid.*

102. Cookson, Clive, 'Cloned body parts, not babies, may get the thumbs up', *Financial Times*, 9 December 1998.

103. Enriquez, J., 'Genomics and the World's Economy', *Science*, Vol. 281, 14 August 1998, p. 925.

104. WO9633276 'Nucleotide sequence of H Influenzae Rd genome, fragments thereof and uses therefore'.

105. Human Genome Sciences, Inc., 'Human Genome Sciences Announces *Haemophilus Influenza* B Vaccine Project', News Release, 28 July 1995.

106. *Ibid.*

107. Erickson, D., 'Microbial Genomics', *Start-Up*, December 1997. See also, Culotta, Elizabeth, 'Science's Breakthroughs of the Year', *Science*, Vol. 278, Number 5346, 19 December 1997, p. 2038.

108. Koonin, E., 'Genomics microbiology: Right on target?', *Nature Biotechnology*, Vol. 16, September 1998, p. 821.

109. Ferry, G., 'The Human Worm', *New Scientist*, 5 December 1998, p. 33.

110. Belkin, L., 'Splice Einstein & Sammy Glick. Add a Little Magellan', *New York Times Magazine*, 23 August 1998.

111. Moukheiber, Z., 'A hail of silver bullets', *Forbes*, 26 January 1998, p. 78.

112. Thompson, Dick, 'Gene Maverick', *Time*, 11 January 1999, p. 55.

113. Sansom, C., 'Unravelling the human genome', *Scrip Magazine*, September 1998, p. 45.

114. Wade, N., 'It's a Three-Legged Race to Decipher the Human Genome', *New York Times*, 23 June 1998.

115. Belkin, L., *op. cit.*

116. PerkinElmer Press Release, 9 May 1998, 'PerkinElmer, Dr J. Craig Venter, and TIGR Announce Formation Of New Genomics', on the internet: http://www.perkin-

elmer.com/press/prc5447.html

117. Wade, N., 'In Genome Race, Government Vows to Finish First', *New York Times*, 15 September 1998.
118. Wade, N., 'It's a Three-Legged Race to Decipher the Human Genome', *New York Times*, 23 June 1998.
119. PR Newswire, 'Company's PCT Publications Now Cover 476 Human Genes', 17 September 1998.
120. Personal communication with John Doll, US Patent and Trademark Office, Biotechnology Patent Division, 22 September 1998.
121. McFarling, Usha Lee, 'The Code War: Biotech Firms Engage in High-Stakes Fight Over Rights to the Human Blueprint', *San Jose Mercury News*, 17 November 1998.
122. Incyte News Release, 'Incyte Reports Year-End Results and 1999 Financial Targets', 3 February 1999. On the internet: http://www.incyte.com
123. Hencke, D., R. Evans and T. Radford, 'Blair (UK) and Clinton (US) Push to Stop Gene Patents', *The Guardian*, 20 September 1999.
124. Kher, U., 'A Man-Made Chromosome', *Discover*, 8 January 1998, p. 40.
125. Ikeno, M. et al., 'Construction of YAC-based mammalian artificial chromosomes', *Nature Biotechnology*, Vol. 16, May 1998, p. 431.
126. Taylor, R., 'Superhumans', *New Scientist*, 3 October 1998.
127. Stix, G., 'Personal Pills: Genetic differences may dictate how drugs are prescribed', *Scientific American*, October 1998.
128. *Ibid.*
129. Sansom, C., *op. cit.*
130. PR Newswire, 'Bristol-Myers Squibb Commissions Largest-Ever Haystack New Automated System; Will Accelerate High-Throughput Drug Discovery', 9 September 1998.
131. Inglis, Julian, 'Traditional Health Systems', *Nature and Resources*, Vol. 30, No. 2, 1994, p. 3.
132. Government of India, Official Press Release, 'Asian Seminar on IPR Issues', 7 October 1998. On the www: http://206.252.12/gov/press/Oct07
133. Anonymous, 'Herbs as drugs', R&D Directions, January/February 1998, p. 58.
134. *Ibid.*
135. Coen, S., 'Fast Forward into the World of Genomics', *Seed World*, September 1998, p. 26.
136. Ratner, M., 'Competition drives agriculture's genomics deals', *Nature Biotechnology*, Vol. 16, September 1998, p. 810.
137. Freiberg, B., 'Top Pioneer Executives Discuss a Completely Changing Seed Industry', *Seed and Crops Digest,* June/July 1999, p. 12.
138. Delta and Pine Land Company, 'Delta and Pine Land Company and the USDA Announce Receipt of Varietal Crop Protection System Patent', 3 March 1998. On the internet: http://biz.yahoo.com/prnews/980303/ms_delta_p_1.html
139. RAFI, 'Traitor Tech: The Terminator's Wider Implications', *RAFI Communiqué,* January/February 1999. On the internet: http://www.rafi.org
140. Subsidiary Body on Scientific, Technical and Technological Advice, 'Consequences of the Use of the New Technology for the Control of Plant Gene Expression for the Conservation and Sustainable Use of Biological Diversity', UNEP/CBD/SBSTTA/4/1/Rev.1, 17 May 1999, prepared for SBSTTA Fourth Meeting, Montreal, 21–25 June 1999, p. 48.
141. RAFI, 'The Terminator Technology', *RAFI Communiqué*, March/April 1998.
142. Freiberg, B., 'Is Delta and Pine Land's Terminator Gene a Billion Dollar Discovery?', *Seeds and Crop Digest*, May/June 1998.
143. *Ibid.*
144. RAFI, 'The Terminator Technology: New Genetic Technology Aims to Prevent Farmers from Saving Seed', *op. cit.*
145. Food and Agriculture Organization of the United Nations, 'The State of the World's

Plant Genetic Resources for Food and Agriculture', (background documentation prepared for the International Technical Conference on Plant Genetic Resources, Leipzig, Germany, 17–23 June 1996), Rome, 1996.

146. Subsidiary Body on Scientific, Technical and Technological Advice, 'Consequences of the Use of the New Technology for the Control of Plant Gene Expression for the Conservation and Sustainable Use of Biological Diversity', UNEP/CBD/SBSTTA/4/ 1/Rev.1, 17 May 1999, prepared for SBSTTA Fourth Meeting, Montreal, 21–25 June 1999, p. 11.

147. Ramsay, J., Unpublished paper distributed by Jonathan Ramsay of Monsanto Europe, Brussels, August 1998, jonathan.ramsay@monsanto.com

148. Collins, H., 'New Technology and Modernizing World Agriculture'. Unpublished paper distributed by Dr Collins at the June 1998 meeting of the FAO Commission on Genetic Resources for Food and Agriculture in Rome.

149. Anonymous, 'India Wary of Terminator', *Nature Biotechnology*, Vol. 16, September 1998.

150. On 30 October 1998 the CGIAR announced: 'The CGIAR will not incorporate into its breeding materials any genetic systems designed to prevent seed germination. This is in recognition of (a) concerns over potential risks of its inadvertent or unintended spread through pollen; (b) the possibilities of sale or exchange of viable seed for planting; (c) the importance of farm-saved seed, particularly to resource-poor farmers; (d) potential negative impacts on genetic diversity; and (e) the importance of farmer selection and breeding for sustainable agriculture.'

151. Letter from Dr D. A. Evans of Zeneca Agrochemicals to Richard Jefferson, CAMBIA, Australia, dated 24 February 1999. The letter was included in documentation provided for SUBSTTA, Fourth Meeting, Montreal, 21–25 June 1999, Item 4.6 of the Provisional Agenda.

152. Dr Gordon Conway, speech to the Monsanto board of directors, 24 June 1999 in Washington, DC.

153. Monsanto Corporation, 'Gene Protection Technologies: A Monsanto Background Statement', April 1999, on the Monsanto web site: http://monsanto.com/monsanto/ terminator/default.htm

154. Open Letter from Monsanto CEO Robert B. Shapiro to Rockefeller Foundation President Gordon Conway and others, 4 October 1999. On the internet: http:// www.monsanto.com/monsanto/gurt/default.htm

155. Becker, H., 'Revolutionizing Hybrid Corn Production', *Agricultural Research*, published by USDA, December 1998.

156. Bicknell, R., and K. Bicknell, 'Who will benefit from apomixis?', *Biotechnology and Development Monitor*, No. 37, March 1999.

157. *Ibid.* The US Department of Agriculture, the International Maize and Wheat Improvement Centre in Mexico (CIMMYT), the French Research Institute for Cooperative Development (ORSTOM), the UK's John Innes Centre, University of California, and Utah State University are among the public research institutes that hold patents related to apomixis technology.

158. For further information about apomixis and the Bellagio Apomixis Declaration, go to: http://billie.harvard.edu/apomixis/

159. Bengtsson, Bo, and Carl-Gustaf Thornström, 'Biodiversity and Future Genetic Policy: A Study of Sweden', ESDAR, Special Report No. 5, The World Bank and Sida, April 1998. See also: Collins, Wanda, and Michel Petit, 'Strategic Issues for National Policy Decisions in Managing Genetic Resources', ESDAR Special Report No. 4, The World Bank, April 1998.

160. UNDP, Human Development Report 1999, p. 60.

161. UNESCO, *Our Creative Diversity:* Report of the World Commission on Culture and Development, 1995, p. 107.

162. UNDP, Human Development Report 1999, p. 58.

163. The Report of the Panel of Experts on Access and Benefit-Sharing, First Meeting, San

Jose, Costa Rica, 4–8 October 1999.

164. IUCN, 'A Guide to the Convention on Biological Diversity', p. 84.

165. The first resolution, C4/89, provides an agreed interpretation which recognized that Plant Breeders' Rights were not necessarily inconsistent with the IU. The second resolution, C5/89 recognized 'Farmers' Rights' and defined them. The third resolution reaffirmed the sovereign rights of nations over their genetic resources and agreed in principle that Farmers' Rights should be implemented through an international fund.

166. FAO, CGRFA-8/99/Rep, Appendix E, Text for Articles 11,12 and 15, established by the Contact Group During the Eighth Regular Session of the Commission.

167. See Resolution 5/89, adopted by the 25th Session of the FAO Conference in November, 1989. This resolution states, *inter alia*, that: 'Farmers' Rights mean rights arising from the past, present and future contributions of farmers in conserving, improving and making available plant genetic resources...'.

168. Commission on Human Rights, Sub-Commission on Prevention of Discrimination and Protection of Minorities, 'The Realization of Economic, Social and Cultural Rights', Updated study on the right to food, submitted by Mr Asbjørn Eide, 28 June 1999, E/CN.4/Sub.2/1999/12.

169. FAO, CGRFA-8/99/Rep, Appendix E. Text for Article 15—Farmers' Rights, established by the Contact Group During the Eighth Regular Session of the Commission.

170. Resolution 3 of the Nairobi Final Act of the Conference for the Adoption of the Agreed Text of the CBD.

171. Food and Agriculture Organization of the United Nations, 'Global Plan of Action for the Conservation and Sustainable Utilization of Plant Genetic Resources for Food and Agriculture and the Leipzig Declaration adopted by the International Technical Conference on Plant Genetic Resources', Leipzig, Germany, 17–23 June 1996.

172. RAFI press release, 'Toward A Global Moratorium on Plant Monopolies', 9 February 1999.

173. Rural Advancement Foundation International and Heritage Seed Curators Australia, 'Plant Breeders' Wrongs: An Inquiry into the Potential for Plant Piracy through the International Intellectual Property Conventions', September 1998. On the internet: http://www.rafi.org

174. In over one-third of all the Australian cases, the Plant Varieties Rights Journal of the PBRO of Australia provides no evidence of breeding activity and there is no other indication that actual breeding took place. This is in contradiction to the 1987 Plant Varieties Act, Section 5 and the Plant Breeders Rights Act of 1994, Section 10, which states: 'Nothing in this Act requires or permits the granting of PBR in a plant variety...unless...the breeding of the plant variety constitutes an invention'. Source: Rural Advancement Foundation International and Heritage Seed Curators Australia, 'Plant Breeders' Wrongs', September 1998. On the internet: http://www.rafi.org

175. For a detailed listing of intellectual property claims (granted and pending), and exclusive licenses on plant varieties under investigation by RAFI and HSCA, go to: http://www.rafi.org/pbr/

176. RAFI press release, 'Plant Breeders' Wrongs Righted in Australia?', 10 November 1999. On the internet: http://www.rafi.org/pr/release24.html

177. New instructions for PBR applicants in Australia were published in the *Plant Variety Rights Journal*, Vol. 11, No. 3, pp. 2–5.

178. Various branches of the Queensland Dept. of Primary Industry were offering a CIMMYT wheat variety, SERI, on a contractual limited propagation basis. CIMMYT took action and the initiative was halted.

179. CIP (International Potato Centre), Annual Report 1998, Lima, Peru, 1999, p. 20.

180. *Ibid.*, pp. 36 and 39.

181. Roberts, Royston M., *Serendipity—Accidental Discoveries in Science,* Wiles & Sons, 1989, p. 54.

182. *New Scientist*, 'Bouncing Back', 26 June 1999, p. 27.

183. 'Fostering Plant Innovation: ASSINSEL Brief on Review of TRIPs 27.3b' at http://

www.worldseed.org/~assinsel/Intprop0.pdf
184. Commission on Human Rights, Sub-Commission on Prevention of Discrimination and Protection of Minorities, 'The Realization of Economic, Social and Cultural Rights', updated study on the right to food, submitted by Mr Asbjørn Eide, 28 June 1999, E/CN.4/Sub.2/1999/12.
185. *The Thammasat Resolution: Building and Strengthening our Sui Generis Rights*, a resolution passed by 45 representatives of indigenous, peasant, non-governmental, academic and government organizations from 19 countries, meeting 1–6 December 1997 near Bangkok, Thailand.
186. Communication from Kenya on behalf of the African Group, 'Preparations for the 1999 Ministerial Conference—The TRIPs Agreement, World Trade Organization', WT/GC/W/302, 6 August 1999. On the internet: http://www.wto.org/wto/ddf/ed/public.html
187. WTO General Council, Preparations for the 1999 Ministerial Conference, 'Proposal on Protection of the Intellectual Property Rights Relating to the Traditional Knowledge of Local and Indigenous Communities', Communication from Bolivia, Colombia, Ecuador, Nicaragua, and Peru, WT/GC/W/36212, October 1999.
188. Letter from David R. Downes, Senior Attorney, Centre for International Environmental Law, to Todd Dickinson, Acting Assistant Secretary of Commerce and Acting Commissioner of Patents and Trademarks, 30 March 1999.
189. Centre for International Environmental Law, CIEL press release, 'US Patent Office Cancels Patent on Sacred 'Ayahuasca' Plant', 4 November 1999.
190. The Final Act Embodying the Results of the Uruguay Round of Multilateral Trade Negotiations includes the Agreement Establishing the World Trade Organization, and, *inter alia*, the following multilateral agreements which are binding on all contracting parties: 1) the Multilateral Agreements on Trade in Goods; 2) the General Agreement on Trade in Services; 3) the Agreement on Trade-Related Aspects of Intellectual Property Rights.
191. UNDP, Human Development Report 1999, p. 67.
192. UNCTAD, 'The TRIPS Agreement and Developing Countries', United Nations, Geneva, 1996.
193. 'The Nairobi Statement', International Conference on Trade-Related Aspects of Intellectual Property Rights and the Convention on Biological Diversity, organized by the African Centre for Technology Studies (ACTS) in conjunction with the United Nations Environment Programme (UNEP), 6–7 February 1999.
194. Two countries are still party to the 1961 Act, revised 1972: Belgium and Spain.
195. UPOV press release, 'The 1991 Act of the International Convention for the Protection of New Varieties of Plants Enters into Force', Geneva, 21 April 1998.
196. ASSINSEL, 'Development of New Plant Varieties and Protection of Intellectual Property', Statement adopted at the Melbourne Congress in June 1999. On the internet at: http://www.worldseed.org
197. See, for example, Ekpere, J.A., 'Alternative to UPOV for the Protection of New Plant Varieties', Paper distributed at the UPOV-WIPO-WTO Joint Regional Workshop on 'The Protection of Plant Varieties under Article 27.3(b) of the TRIPS Agreement', Nairobi, 6–7 May 1999. See also, Sahai, Suman, 'Protection of New Plant Varieties: A Developing Country Alternative', Economic and Political Weekly, Vol. XXXIV, Nos. 10 & 11, March 1999.
198. In fact, there is no specific mention of farm-saved seed in the 1978 Act of the UPOV Convention. Article 5(1) on rights protected and scope of protection provides: '(1) The effect of the right granted to the breeder is that his prior authorization shall be required for: the production for purposes of commercial marketing; the offering for sale; and the marketing of the reproductive or vegetative propagating material, as such, of the variety.' Because only commercial marketing is forbidden, production for other purposes is allowed.
199. Article 15(2) of the 1991 Act: '[*Optional exception*] Notwithstanding Article 14, each

Contracting Party may, within reasonable limits and subject to the safeguarding of the legitimate interests of the breeder, restrict the breeder's right in relation to any variety in order to permit farmers to use for propagating purposes, on their own holdings, the product of the harvest which they have obtained by planting, on their own holdings, the protected variety or a variety covered by Article 14(5)(*a*)(i) or (ii).'

200. Article 5(1)(i) of the 1991 Act: 'Compulsory exceptions: the breeders' rights shall not extend to acts done privately for non-commercial purposes.'

201. This was a voluntary feature of the 1978 Act, Article 5(4).

202. See Leskien, Dan, & Michael Flitner, 'Intellectual Property Rights and Plant Genetic Resources: Options for a *Sui Generis* System', *Issues in Genetic Resources*, No. 6, Rome, 1997.

203. For further background, see 'Beyond UPOV: Examples of developing countries preparing non-UPOV *sui generis* plant variety protection schemes for compliance with TRIPs', GRAIN, July 1999. Available on the Internet: http://www.grain.org/publications/reports/nonupov.htm

204. Ekpere, J.A., 'Alternative to UPOV for the Protection of New Plant Varieties', paper distributed at the UPOV-WIPO-WTO Joint Regional Workshop on 'The Protection of Plant Varieties under Article 27.3(b) of the TRIPs Agreement', Nairobi, 6–7 May 1999.

205. Rivera, V.S., and P.M. Cordero, 'Costa Rica's Biodiversity Law: Sharing the Process', paper prepared for the workshop on 'Biodiversity Conservation and Intellectual Property Regime', organized by the Research and Information System for the Non-Aligned and Other Developing Countries (RIS) with the World Conservation Union, New Delhi, India, 29–31 January 1999.

206. International Centre for Trade & Sustainable Development, 'TRIPs Council discusses plant patenting', *Bridges Weekly Trade News Digest*, Vol. 3, No. 15/16, Geneva, 26 April 1999.

207. Ekpere, J.A., *op.cit.* See also 'TRIPs vs. Biodiversity: What to do with the 1999 review of Article 27.3(b)', GRAIN, May 1999, URL: http://www.grain.org/publications/reorts/tripsmay99.htm

208. Anonymous, 'US Sees No TRIPs Negotiations at Seattle, Focuses on Implementation', *Inside US Trade*, Vol. 17, No. 31, 6 August 1999.

209. International Centre for Trade and Sustainable Development, 'North–South Divide Splits TRIPs Council', *Bridges Weekly Trade News Digest*, Vol. 3, No. 42, 25 October 1999.

210. ISNAR, 'The Use of Proprietary Biotechnology Research Inputs at Selected CGIAR Centres', draft version, 29 January 1998.

211. Reeves, T.G., 1997, 'Apomixis Research: Biotechnology for the Resource-Poor—Some Ethical And Equity Considerations'.

212. CGIAR, 'Report of the CGIAR Expert Panel on Proprietary Science and Technology', Mid-Term Meeting, 25–29 May 1998, Brasilia, Brazil.

213. Leskien, D., 'The European Patent Directive on biotechnology', *Biotechnology and Development Monitor*, No. 36, September/December 1998, pp. 16–19.

214. It should be noted that the European Patent Convention has more member states than the EU. EPC member states include all EU member states plus Cyprus, Liechtenstein, Monaco and Switzerland.

215. For further background on the European patent system, see: Leskien, D., *op. cit.*, pp. 16–19. Leskien notes that the Patent Directives, understanding of 'plant varieties' is incompatible with the EPC. However, the EPC does not prevent its member states from granting patents for inventions that are excluded from patentability under the EPC.

216. In June 1998 the Swiss held a national referendum on gene patenting. The referendum, which would have outlawed the patenting of plants and animals, was defeated. (Switzerland is not a member of the EU, but it is a member of the European Patent Convention.)

217. *Nature*, 10 December 1998.

218. The relevant provisions are clauses in the Preamble, reading as follows:
     '(26) Whereas if an invention is based on biological material of human origin or if it uses such material, where a patent application is filed, the person from whose body the material is taken must have had an opportunity of expressing free and informed consent thereto, in accordance with national law;
     (27) Whereas if an invention is based on biological material of plant or animal origin or if it uses such material, the patent application should, where appropriate, include information on the geographical origin of such material, if known; whereas this is without prejudice to the processing of patent applications or the validity of rights arising from granted patents.'
219. In June 1999, private and public plant breeders from 31 industrialized and developing countries representing more than 1 000 seed companies met in Melbourne at a meeting of ASSINSEL to define their position on IP. The ASSINSEL members adopted a motion stating that 'when a commercially available plant variety protected by plant variety protection contains patented traits, it should remain freely available for further breeding, according to the breeders' exception provided for in the UPOV and UPOV-like systems.'
220. Cohen, S., 'Unravelling code of genetic patents', *The Lawyer*, 29 September 1998.
221. There are three separate intellectual property systems covering plants in the United States: 1) the 1930 Plant Patent Act (PPA) for new varieties of asexually reproduced plants; 2) the 1970 Plant Variety Protection Act for sexually propagated plant varieties; 3) US utility (protection extended to plants by the US Patent and Trademark Office in 1985).
222. The US government argues that the USA concludes regional and bilateral agreements that neither create closed regional trading blocs nor erect barriers to non-participants. Neither of these types of programs would be permitted under TRIPs.
223. Anonymous, 'Japan, US, EU to integrate patent screening systems', Japan Economic News Wire, 14 November 1997.